The Reform of the International System of Units (SI)

T0187762

Systems of units still fail to attract the philosophical attention they deserve, but this could change with the current reform of the International System of Units (SI). Most of the SI base units will henceforth be based on certain laws of nature and a choice of fundamental constants whose values will be frozen. The theoretical, experimental and institutional work required to implement the reform highlights the entanglement of scientific, technological and social features in the scientific enterprise, while it also invites a philosophical inquiry that promises to overcome the tensions that have long affected science studies.

Nadine de Courtenay is Associate Professor of History and Philosophy of Science at the Paris Diderot University & Laboratoire SPHERE, France.

Olivier Darrigol is Research Director at CNRS & Laboratoire SPHERE, France.

Oliver Schlaudt is Associate Professor of Philosophy of Science at Heidelberg University, Germany.

History and Philosophy of Technoscience

Series Editor: Alfred Nordmann

For more information about this series, please visit: www.routledge.com

The Reform of the International System of Units (SI)

Philosophical, Historical and Sociological Issues

Edited by Nadine de Courtenay,
Olivier Darrigol, and Oliver Schlaudt

Routledge
Taylor & Francis Group

LONDON AND NEW YORK

First published 2019
by Routledge
2 Park Square, Milton Park, Abingdon, Oxon OX14 4RN

and by Routledge
52 Vanderbilt Avenue, New York, NY 10017

First issued in paperback 2020

Routledge is an imprint of the Taylor & Francis Group, an informa business

British Library Cataloguing-in-Publication Data
A catalogue record for this book is available from the British Library

Library of Congress Cataloging-in-Publication Data
Names: Courtenay, Nadine de, 1961– editor. | Darrigol, Olivier,
    editor. | Schlaudt, Oliver, editor.
Title: The reform of the International System of Units (SI) :
    philosophical, historical, and sociological issues / edited by
    Nadine de Courtenay, Olivier Darrigol, and Oliver Schlaudt.
Description: Abingdon, Oxon ; New York, NY : Routledge, 2019. |
    Series: History and philosophy of technoscience ; 15 | Includes
    bibliographical references.
Identifiers: LCCN 2018045789 (print) | LCCN 2018049894
    (ebook) | ISBN 9781351048989 (eBook) | ISBN
    9781138483859 (hardback : alk. paper)
Subjects: LCSH: Metric system. | Measurement—History.
Classification: LCC QC91 (ebook) | LCC QC91 .R4354 2019
    (print) | DDC 530.8/1209—dc23
LC record available at https://lccn.loc.gov/2018045789

ISBN 13: 978-0-367-66260-8 (pbk)
ISBN 13: 978-1-138-48385-9 (hbk)

Typeset in Galliard
by Apex CoVantage, LLC

Contents

Figures and tables

Figures

Tables

Contributors

Christian Bordé is Emeritus Director of Research at CNRS, member of the Institut Laboratoire de Physique des Lasers, UMR 7538 CNRS, Université Paris Nord; LNE-SYRTE, UMR 8630 CNRS, Observatoire de Paris, France.

Nadine de Courtenay is Associate Professor of History and Philosophy of Science at Paris Diderot University and UMR SPHERE, France.

Olivier Darrigol is Director of Research at CNRS, UMR SPHERE, France.

Alessandro Giordani is Associate Professor in the Department of Philosophy at the Università Cattolica, Milan, Italy.

Ingvar Johansson is Professor Emeritus in the Department of Philosophy at Umeå University, Sweden.

Jean-Marc Lévy-Leblond is Professor Emeritus in the Department of Physics, Université de Nice-Sophia Antipolis, France.

Luca Mari is Professor of Measurement Science at Università Cattaneo – LIUC, Castellanza, Italy.

Martin J. T. Milton is Director of the International Bureau of Weights and Measures, Pavillon de Breteuil, Sèvres, France.

Terry Quinn is Honorary Director of the International Bureau of Weights and Measures, Pavillon de Breteuil, Sèvres, France, and Fellow of the Royal Society.

Sally Riordan is Senior Research Fellow at the Education Observatory, University of Wolverhampton, England.

Oliver Schlaudt is Associate Professor of Philosophy of Science in the Department of Philosophy at Heidelberg University, Germany.

Susan G. Sterrett is Curtis D. Gridley Distinguished Professor in the History and Philosophy of Science in the Department of Philosophy at Wichita State University, Wichita, Kansas, USA.

Contributors

Introduction

Nadine de Courtenay, Olivier Darrigol
and Oliver Schlaudt

Measurement is ubiquitous. It has been part and parcel of everyday activities in commerce, manufacturing and administration since ancient times, and it is the hallmark of our most sophisticated contemporary scientific enterprises. As a result of the entwinement of science and technology, metrology, the science of measurement, now pervades the very infrastructure of our society. Moreover, as society becomes more globalized there is a growing need for how we measure to be reviewed; tighter, worldwide comparability of measurements and standards is required.

The significance of measurement, however, does not figure highly in the current agenda for philosophy of science. Measurement has suffered even more than experiment from neglect, as philosophers have turned their attention to theoretical concerns. Measurement featured prominently in the analyses of the logical empiricists, but by the second half of the twentieth century the focus had switched. The purely formal developments of the representational theory of measurement and the conceptual investigations of metaphysics usurped research into more practical measurement issues, while philosophers ceased altogether to engage with epistemological questions related to measurement. It is only in very recent times that signs of a revival of interest in measurement have emerged. In the twenty-first century there have been a number of publications and international conferences that testify to a growing interest in concrete measurement practices and the historical, experimental, material and social dimensions of measurement (Chang, 2004; Boumans, Hon, & Petersen, 2014; Tal, 2015; Schlaudt & Huber, 2015; Mössner & Nordmann, 2017; Mitchell, Tal, & Chang, 2017).

Despite this revived interest in the material and social underpinnings of measurement, the study of systems of units still fails to attract the attention it deserves. Some historians, interested in the changes of units and standards over time, as well as sociologists, attentive to the role of standardization and to settling arguments over measurement results, have turned their attention to metrology and the subject of units and standards. Few philosophers, however, have shown much interest in the area. This, perhaps, could change with the reform of the International System of Units (SI) proposed in 2006 and recommended by the Consultative Committee on Units (CCU) in September 2017. The SI reform that has been endorsed by the General Conference on Weights and Measures (CGPM) on 16 November 2018 should indeed prompt debates about the philosophical ramifications of the construction of a system of units.

In the earlier system, the base units rest on a heterogeneous set of definitions. The kilogram is defined in terms of a material artefact, the second is related to a specific atomic microwave transition in caesium atoms, whilst the definition of the metre is based on the second and on the fixed value of a fundamental constant, the velocity of light. The new system moves to redefine most of the SI base units on the model of the metre: the definitions of the kilogram, the ampere, the kelvin and the mole are based on certain laws of nature and a choice of fundamental constants whose values are frozen, as they have been measured with an uncertainty sufficiently small to allow the revision of the system without introducing inconsistencies. Such a system of units, based on fundamental constants (velocity of light, Planck's constant, Boltzmann's constant, Avogadro's constant), is expected to better fulfil the demands of invariability, permanence, accessibility and universality that are required of a system of units. A considerable amount of theoretical, experimental and institutional work has been carried out during more than a decade at the international level to make this reform possible.

Devising a measurement system based on invariant quantities of nature has been a goal ever since the inception of the metric system. The building of a system of units based on fundamental constants has become possible today thanks, on the one hand, to the theoretical and experimental developments of the twentieth century, which enable us to measure the fundamental constants with unprecedented accuracy, and, on the other hand, to the discovery of quantum-based measurement principles (laser spectroscopy, Josephson effect, quantum Hall effect) that establish a bridge between the microscopic level at which the determination of the fundamental constants takes place and the macroscopic scale at which common measurements are performed.

The SI reform is thus connected with a major transformation of metrology. Going beyond its traditional task of maintaining and disseminating standards, metrology has now become inseparable from fundamental theoretical and experimental issues. Whereas units defined by prototypes were disconnected from laws of nature, units related to fundamental constants become inseparable from deep theoretical considerations and sophisticated experimental realizations involving atomic and solid-state quantum phenomena. The reform is also triggering a decisive re-organization of the architecture of the dissemination process and therefore, profound institutional changes at the international level.

The SI reform thus makes particularly salient the entanglement of the scientific, technological and social features that come into play in the scientific enterprise. In addition to profound theoretical questions, the construction of the New SI involves normative activities governing the communicability and comparability of scientific results, and it depends on a large number of interactions that are coordinated by institutions and encompass epistemic as well as material components. The study of the SI reform thus seems to offer an opportunity to bridge the gulf that has divided philosophical, historical and sociological approaches of science during the last decades of the previous century. This book aspires to promote this opportunity by presenting the SI reform to the philosophical community and offering a number of philosophical reflections that the SI reform has already inspired. All contributions were written before the decision taken by the CGPM on 16 November 2018.

Three metrologists have been invited to present the reform. They approach the subject from different perspectives and thus help to construct a comprehensive view of the enterprise.

Terry Quinn, emeritus director of the International Bureau of Weights and Measures (BIPM) from 1988 to 2003, sets the stage by positioning the SI revision on the background of the manifold transformations that have accompanied the development of metrology from its eighteenth-century beginnings to the frontiers of today's science and technology. He expounds the key advances that make the reform possible at present and the conditions that must be met in order to carry it out effectively. His account of the abandonment of standards based on material artefacts in favour of new definitions of units anchored in fundamental physics stresses the need for accuracy over and above the mere demands of comparability and reproducibility. The impact of the new definitions on the practical realization of the base units also brings him to tackle institutional issues related to the role of the National Metrology Institutes, whose purpose is to guarantee to the industrial nations worldwide access to comparable measurement standards.

In his chapter, Martin J. T. Milton, director of the BIPM, discusses the motivations behind the "New SI". He goes back to the seminal articles of 2005 and 2006 by Mills et al., as well as to several other publications that have advocated change to the SI, in order to evaluate the respective import of the technical and the conceptual reasons articulated in favour of the reform. Milton's analysis shows that the pragmatic arguments, principally related to mass and electrical units, involve a subtle combination of practical advantages and disadvantages that is not conducive to a widespread support of the reform since it affects users working in different fields in quite opposite ways. He then explores the more general, conceptual arguments, such as the idea that base units should be defined in terms of "invariants of nature" and the claim that the new definitions would help to further our understanding of theoretical and quantum physics. He finally insists on the importance of a group of arguments related to the realization of the base units: the new definitions promote the universality, freedom and equality of access to primary realizations of the base units.

In his chapter, French physicist Christian Bordé, four times president of the General Conference on Weights and Measures, leads the reader even further beyond pragmatic justifications of the New SI. Bordé sets out to reveal the deepest theoretical underpinnings of the reform using an original, in-depth analysis of the redefinition of the unit of mass. In his analysis, Bordé demonstrates how the redefinition rests on a link between mass and proper time that involves relativistic quantum theory, which can be investigated using atomic interferometry. Bordé also addresses the crucial question as to whether such a link is still significant at the macroscopic level of the kilogram. The definitions of the other units of the SI are discussed with equal attention to fundamental and experimental issues. The chapter culminates in setting the reform within an overarching framework based on the development of a geometry in five dimensions.

The five other chapters of this book discuss the SI reform from a philosophical point of view.

The philosopher Alessandro Giordani and the philosopher-physicist Luca Mari, who is also a member of the Joint Committee for Guides in Metrology (JCGM) currently working on the revision of the *International Vocabulary of Metrology* (*VIM*), carefully introduce the notions of systems of quantities and systems of units and discuss strategies for defining units. They argue that "Global Constant Definitions" that provide a simultaneous definition of all units in terms of a set of natural constants best satisfy the desiderata of an ideal system of units (such as stability, intersubjectivity, minimal uncertainty). Furthermore, Giordani and Mari tackle the question of how systems of units can be changed in response to new insights into the underlying quantities. They show that it is possible to design a Global Constant Definition in such a way that changes stemming from technological development can simply be dealt with by adapting a set of scale factors, and only changes due to new scientific insights would necessitate a revision of the underlying set of natural units.

In her chapter, the American philosopher Susan G. Sterrett urges philosophy of science to pay more attention to the issue of the coherence of systems of units. Indeed, in the course of the reform the units of the SI become more interrelated, and, as a consequence, issues of coherence gain in importance. A special feature of the New SI is that it exhibits a definition free of the distinction of base and derived units. Thus, in the New SI, the constraint of coherence is implemented differently than in the existing international system, where it relies on the formalism of base units and derived units. Sterrett examines to what extent the distinction between base and derived units can really be dispensed with in a system of coherent units. This leads her to unravel the relations between units and quantities, showing that the establishment of a system of units is a "holistic affair" based on the statement of quantity equations. She also straightens out some mistaken ideas that are often held about the notion of dimension.

The reform of the SI reveals constants of nature in a new light. The French physicist Jean-Marc Lévy-Leblond provides conceptual insights into the nature of physical constants by delivering an updated version of his classical conceptual analysis of physical constants from 1977. In particular he suggests a classification of natural constants, distinguishing between three types of constants, A, B and C: physical properties of particular objects (e.g. the mass of an elementary particle), constants characterizing classes of phenomena (e.g. coupling constants) and universal constants such as h or c as they figure in the most fundamental theories, where, according to Lévy-Leblond, they can be regarded as "theory synthesizers". Relying on historical examples, Lévy-Leblond examines the role constants play in various strands of physics and shows that they move "from one type to another when our physical knowledge increases". The velocity of light, for instance, started as a mere physical property of light, then became a characteristic of all electromagnetic phenomena following the publication of Maxwell's theory of electromagnetism, and finally it revealed its character as a universal constant in the formulation of the Lorentz transformation. Summing up his main thesis, Lévy-Leblond contends: "Only by studying the conditions for the appearance, or disappearance, of physical constants can we understand their nature".

The philosopher Sally Riordan, now at the University of Wolverhampton, denounces the philosophers' neglect of metrology and offers a view of the physical sciences in which theory, experiment and metrology play equally important roles and permeate each other. She first defines and criticizes the received "conventionalist account of scientific standards" according to which the choice of a standard is entirely a matter of convention, guided by human values and not by nature. To this account she opposes a "metrological account" based on the actual practice of metrologists. The latter account is compatible with a concept of metrological progress, and it implies a considerable entanglement of the metrological, experimental and theoretical aspects of physics. Taking the example of the old and new definitions of the kilogram in the SI system, Riordan first argues that theoretical commitments may largely motivate a change of standard. Second, she observes that the same device (the watt balance for instance) may be regarded either as a means to measure a constant (Planck's constant) or as a calibration device (for a mass standard), the decision between these two options belonging to the metrologists; this blurs the frontier between experimentation and metrology. Third, she argues that the full definition of a standard should include the *mise en pratique*, which implies both experimental and theoretical components. All through her chapter, she denounces the incompatibility of various philosophical positions with this "metrological account", and she propounds a "metrological operationalism" in which physical concepts are at least partially defined by the collection of methods that realize them, and they can be combined with a "metrological realism" in which nature can favour some definitions of standards over others.

In his contribution, the Swedish philosopher Ingvar Johansson addresses the question of the constancy of units and the concomitant difficulties in properly defining the base units in the SI system. Some of these difficulties, such as the necessary theoretical embedding of basic quantities or the "surround knowledge" needed in the concrete implementation of a defined unit, he regards as inevitable and manageable. Others he regards as consequences of misunderstandings or wrong choices. In his opinion, for instance, the distinction between base units and derived units in the SI system falls into this latter category because the true logic of the definition of a given unit through a universal constant (and other units) is to regard the universal constant as a base unit and the given unit as a derived unit.[1] Johansson also deplores the new definition of the kilogram through the Planck constant (and other constants and units), for it relies on a constant (h) and on a relation ($h\nu = mc^2$) that have no direct empirical or material realization,[2] and because in his opinion the quantum-relativist framework needed to define the latter relation is not sufficiently established.[3] Lastly, Johansson notes that in the drafts for the new system the issue of how to test the constancy of the fundamental constant used in the definition is not addressed.

Notes

1 On the basis of other contributions to this volume, it could be argued that the metrologists' choice is nonetheless justified because it remains acceptable as a convention, because it allows the usual dimensional analysis and because it reflects a natural hierarchy in the definition of physical concepts.
2 For a different view, see Bordé's chapter in this volume, Section 6.

3 It could still be argued that no direct empirical materialization is needed as long as the constants are basic components of extremely well corroborated theories (no full quantum-relativistic theory of everything is needed for this purpose) and as long as the theories used in the concrete realization of the definition of the kilogram (through the Watt Balance, the Josephson effect and the quantum Hall effect) are sufficiently robust. Of course, no such realization is entirely foolproof, but there are tests of inner consistency as well as possible comparisons with alternative definitions (such as the silicium-based definition).

References

Boumans, M., Hon, G., & Petersen, A. C. (2014). *Error and uncertainty in scientific practice*. London: Pickering & Chatto.

Chang, H. (2004). *Inventing temperature: Measurement and scientific progress*. Oxford: Oxford University Press.

Mitchell, D. J., Tal, E., & Chang, H. (2017). *The making of measurement*. International Conference, University of Cambridge, 23–25 July 2015, special issue of *Studies in History and Philosophy of Science*, Part A, vols. 65–66.

Mössner, N., & Nordmann, A. (Eds.). (2017). *Reasoning in measurement*. London: Routledge.

Schlaudt, O., & Huber, L. (Eds.). (2015). *Standardization in measurement*. London: Pickering & Chatto.

Tal, E. (2015). Measurement in science. In E. Zalta (Ed.), *Stanford encyclopedia of philosophy* (Summer 2015 ed.). Retrieved from http://plato.stanford.edu/entries/measurement-science.

1 The origins of the Metre Convention, the SI and the development of modern metrology[1]

Terry Quinn

1 Introduction

Few people have any idea of the extensive networks that exist today to ensure accuracy and reliability in measurement. The subject is known as international metrology. Accurate and reliable measurements are essential for almost all aspects of modern life including, for example, international trade, high-technology manufacturing, human health and safety, protection of the environment, global climate studies and the basic science that underpins it all. To meet these multitude requirements for measurement there now exists an International System of Units (SI) that is uniform and accessible worldwide through the work of national metrology institutes. These exist in all the industrialized states of the world and cooperate with each other under the auspices of an intergovernmental treaty signed in 1875, the Metre Convention. The SI is based on our present best theoretical description of nature so that SI units are stable in the long term, internally self-consistent and practically realizable. However, metrology continuously evolves to meet changing needs for measurement and to take account of advances in science and technology. The notable event in this respect was the redefinition of the SI in November 2018 of a new way of defining the SI be based on a set of seven defining constants, drawn from the fundamental constants of physics and other constants of nature.

Neither modern metrology nor today's needs for accurate and reliable metrology appeared suddenly; both are the result of more than 200 years of development. The origin of today's metrology can be traced to two events that took place over a period straddling the end of the eighteenth and beginning of the nineteenth centuries: the first was the creation and implementation in France of the decimal metric system; the second was the development of mass production using interchangeable parts. At the time these two events were not linked, although there is strong evidence that the latter also had its beginnings in France. Nevertheless, the metric system was not created in order to facilitate the production of engineered products, and the early development of mass production did not in any way rely upon the new units of measurement. The origins of the metric system sprang first from attempts to unify and bring some order to the confusion created by the multitude of units used in France in local trade and then embrace the grand idea of producing a set of units that were in some way natural or funda-mental and unrelated to material objects. The development of mass production, on the other hand, was related to the need to produce as many guns as possible in the shortest time and to man's innate urge to maximize profits in so doing!

As we shall see, however, over the past two centuries these two disparate threads have come together. We can now expand the meaning of the term "interchange-able parts" to encompass not only the real interchangeability of components of high-technology manufacturing but also the worldwide comparability of a great diversity of measurements made in almost all aspects of our daily life. Added to this are the much more recent requirements related to human health and safety in its broadest sense. All of these now depend upon a system of measurement that is itself worldwide and based upon a set of units that can be assured to be universal and constant in time and not linked to any one country or place, i.e. as far as pos-sible based on the fundamental constants of nature.

As is well known, the metric system took some time to become established in France; people everywhere have a natural resistance to change, particularly in respect of such basic things as the units in which they transact their everyday busi-ness. It was not until 1840 that the metric system in France became the sole legal system of measurement, although by that time it had been taken up in a number of other European countries. Despite early interest by Sir John Riggs Miller, a British Member of Parliament in the 1780s, and Thomas Jefferson, at that time American Minister to France, neither Great Britain nor America adopted the metric system at the end of the eighteenth century. The American Congress took little notice of Jefferson's proposals when he was Secretary of State to George Washington in the early 1790s, and the British Parliament let the matter drop when Riggs Miller lost his seat at a by-election. Britain went on to bring in a new weights and measures law defining new standards for the yard and the imperial pound in 1824. There were, however, serious attempts to introduce the metric system in Britain and the British Empire during the first decade of the twentieth century. Despite strong support from most of the colonies and many quarters in England these failed, ultimately because of the strong opposition of certain manufacturing trades, who were opposed to the heavy financial costs of changing patterns, draw-ings and machine tools. In other words, the proposals had in one sense come too late. By that time, manufacturing industries had become completely locked into the national standards – the key one being of course length, with the inch as the reference for all engineering tools and designs.

The development of mass production of engineered goods, the precursor of modern high-technology manufacturing, seems to have started in France in 1778. Honoré Blanc, at the time the superintendent of the Royal Ordnance factory at St. Etienne, attempted to introduce a system of production based on pre-constructed filing jigs that could be used by unskilled labour to produce precision parts for the flint-locks of muskets. A hierarchy of standard jigs was established and particular care was attached to the accuracy of the screws and nuts. He managed to produce some 200 locks made from interchangeable parts. Overall, however, the attempt to extend this to other plants failed, and the whole enterprise was abandoned due to opposition from skilled workers who saw their livelihood threatened. The credit for the successful implementation of the first mass production using interchangeable parts is usually given, however, to Eli Whitney, who obtained a contract from the American government in 1798 to produce 10,000 muskets within a period of two years. Although he failed to meet the delivery date (by some ten years) and the interchangeability of the parts was limited, it marked the beginning of large-scale manufacture in the USA, not only of muskets but also of other manufactured goods progressively adapted to the principle of interchangeability of parts. The need very quickly appeared for local standards and a well-established hierarchy of references in each factory. Note that interchangeability could only be attained within each factory; today we strive for worldwide uniformity and interchangeability. This is a difference of degree not a difference of kind, as we shall see.

The rapid development of manufacturing technology during the first half of the nineteenth century was accompanied by, and in fact could hardly have taken place without, a corresponding development in the design and manufacture of measuring machines, standardization of screw threads and indeed such basic things as engineering flat surfaces and straight edges, all of which are essential for precision manufacturing on a large scale. Among the famous names involved were Henry Maudsley, who made what is probably the first accurate measuring machine, which he called his Lord Chancellor (now in the Science Museum, London), and Joseph Whitworth, who was trained by Maudsley. Whitworth is credited with developing, while working for Maudsley, the technique of making a flat surface by successively scraping off the high spots from three flat surfaces, one against each other. In due course, Whitworth was able to make steel plates sufficiently flat enough so that they would stick together by molecular adhesion. He then went on to produce many measuring machines and introduced his system of standard screw threads. By the middle of the nineteenth century, engineering metrology had reached a high level with widely available measuring machines that could measure to 0.0001 inch, with corresponding flat surfaces and straight edges also at the disposal of engineering works. Added to these was the codification of the principles of engineering design that allowed rigid structures to be made with well-fitting components connected together so that linear and circular movements could be obtained. All of this comes under the name of kinematic design. In the 1840s, the principles of engineering design were even beginning to be taught at Cambridge University by Robert Willis, who is thought to have been the person from whom James Clark Maxwell and William Thomson learnt their principles of mechanisms and engineering design.

The next major advance in engineering metrology was made by Carl Eduard Johansson, who in the last decade of the nineteenth century invented the techniques for making accurate gauge blocks by hand lapping using a domestic sewing machine. He made sets of 102 gauges, each having an accuracy of 1 μm. Standards of length in the range from 1 mm to 201 mm, and an accuracy better than 10 μm could be obtained by wringing together combinations of two or more individual gauges.

The stage was thus set for the development of modern metrology.

2   The Metre Convention of 1875: the metric system becomes truly international

While all these advances were being made in engineering metrology and industrial production, there had been no changes in the standards of length and mass established at the time of the French Revolution. Moves towards a formal international adoption of the metric system did not emerge until the middle of the nineteenth century. The Great Exhibition of 1851 held in London was the first of the Great Exhibitions that took place during the second half of the nineteenth century bringing together manufactured products from all over the world. At the 1851 Exhibition, the great advances made in mechanical engineering and in mass production were evident for all to see. Joseph Whitworth exhibited his "millionth" machine, a measuring machine purporting to be able to measure to one millionth of an inch. It was clear that although great advances were being made, there was still a great disparity in units of measurements, not just in length and mass but in many other areas as well.

At the 1855 Exhibition held in Paris, formal moves began, with a view to establishing a worldwide agreement on units of measurement based on the decimal metric system. This was strongly supported by British scientists, despite the fact that the metric system was not used in Britain. The *Commissaires* and members of the jury judging the exhibits – led by Professor Leone Levi, a Fellow of the Royal Society of London – made a formal request at the closure of the Exhibition to Governments there represented for the establishment of a worldwide system of weights and measures based on the decimal metric system. This was supported at about the same time by a request from the Society of Arts and Manufacturing in London to the Treasury for the introduction of the metric system in Britain and the Empire. At the first international statistical Congress held in Paris, also in 1855, James Yates, also a Fellow of the Royal Society, proposed the creation of an international association for the adoption of the metric system. In 1864 the use of the metric system became legal in Britain, and in 1868 it was adopted in Germany. At the Great Exhibition held in Paris in 1867, a Committee for Weights and Measures and Money was formed.

In the same year two other important calls for action were made, and these were the ones that precipitated the French Government to take action. Two particular recommendations of a meeting of European geodesists in Berlin, organized by the newly formed International Association for Geodesy, attracted the attention of the French Government. The first was a call for the construction of a new European standard of the Metre that would be better adapted to modern requirements than was the original Metre of the Archives, which was considered inaccessible, being held in the Archives of France, and the second was for the creation of a European International Bureau of Weights and Measures. These were supported by a separate call from the Academy of Science of Saint Petersburg, which raised the question as to what exactly was the definition of the Metre – was it one ten millionth of the distance from the pole to the equator or was it the length of the Metre of the Archives? The French Bureau de Longitude formally transmitted both of these requests to the French Government and then with representatives of the Academy of Science of Paris and later the Academy of Saint Petersburg made requests to the French Government for the creation of an international commission to oversee the construction of new international metric standards, with a view to thereby make the metric system truly international.

In response, an International Metre Commission was created in 1870 by the French Government. After meetings in 1870 – a bad time to have an international meeting in Paris – and later in 1872, a diplomatic Conference took place in Paris in 1875 at which the Metre Convention was drawn up and signed by 17 Member States. This created an international, permanent scientific institute for metrology, the International Bureau of Weights and Measures, where new international prototypes of the Metre and Kilogram would be deposited for the use of all contracting governments of the Convention. At the Conference there was a strong division of opinion about the purpose of the International Bureau. There were those who considered that it should be a permanent scientific institute charged with carrying out scientific work related to metrology, notably the diffusion and improvement of the metric system. This view was led principally be Professor W. Foerster from Germany. Others considered that after the new prototypes had been constructed the international institute should serve simply as a depository of the metric prototypes to be made available to representatives of member governments on the rare occasions when it became necessary. This view was led by Professor Johannes Bosscha from the Netherlands.

This difference of opinion reflected very clearly the different views being expressed in many European countries at the time as to who should finance and organize scientific research. The enormous growth in science had been led almost exclusively in universities or through independent aristocratic researchers. However, its evident consequences for industrial development and international trade, matters in which governments take a close interest, had led to many requests for government financial support. On the one hand, the university researchers who needed money for the increasingly expensive laboratories did not want government direction, and on the other, the governments, which were almost all liberal and *laissez-faire* in persuasion, did not feel the need to spend taxpayers' money on activities whose immediate outcomes, by their very nature, were unpredictable. The proposed creation of an international laboratory was considered by some as going too far, and that research on measurement standards could only be carried out in universities free of government control. In the end, however, the outcome was a large majority in favour of the new international institute being permanent and scientific, and thus it was created in 1875 and given the name *Bureau International des Poids et Measures* (BIPM). In English it is known as the International Bureau of Weights and Measures. Among the main supporters of this view was Professor Jean-Baptiste Dumas, representing the view of France. Not many years later, its fifth director, Charles Edouard Guillaume, was awarded the 1920 Nobel Prize for physics for his discovery and development of the low thermal expansion alloys of nickel known as Invar and Elinvar. The institute created in 1875 continues today as a scientific institute, although in recent decades it has increased its activity in the international coordination of metrology.

The basic organizational structure created by the Convention (see Figure 1.1) remains as it was created in 1875. A second Convention in 1921 extended the scope of the Convention and made minor modifications related principally to financial arrangements. The purpose of the Convention, as stated in the preamble to the Convention of 1875, was written in the following way:

> The High Contracting Parties (List of names of Heads of States) . . . wishing to assure the international unification and perfection of the metric system have resolved to conclude a Convention to this effect and have nominated their Plenipotentiaries, namely (List of their names) . . . who, after having communicated their full powers, found to be in good and due form, have decided upon the following arrangements:

Article 1

> The High Contracting Parties undertake to create and maintain, at their common expense, a scientific and permanent *International Bureau of Weights and Measures* with its seat in Paris.

Article 2

> The French Government will take the necessary steps to facilitate the acquisition or, if the case arises, the construction of a building specifically dedicated to this purpose, in accordance with the conditions set out in the Regulations annexed to the present Convention.

Article 3

> The International Bureau shall operate under the exclusive direction and supervision of an *International Committee for Weights and Measures*, itself placed under the authority of a *General Conference on Weights and Measures*, consisting of the delegates of all the contracting Governments.[2]

These first three Articles give the essential structure of the Convention. The 12 Articles of the Convention that follow and the 22 Articles of the accompanying Regulations, which have the same force as the Articles of the Convention, established what was the first scientific intergovernmental organization. The Convention set out the mission of the organization and created the three organs to carry it out, shown in italics in Articles 1 and 3. In addition to the BIPM, there is the General Conference on Weights and Measures (CGPM, acronym of the name in French: *Conférence Général des Poids et Mesures*), which now meets in Paris every four years, and the International Committee for Weights and Measures (CIPM, *Comité International des Poids et Mesures*), which now meets at least annually. It is made up of 18 individuals, each from a different Member State and elected or

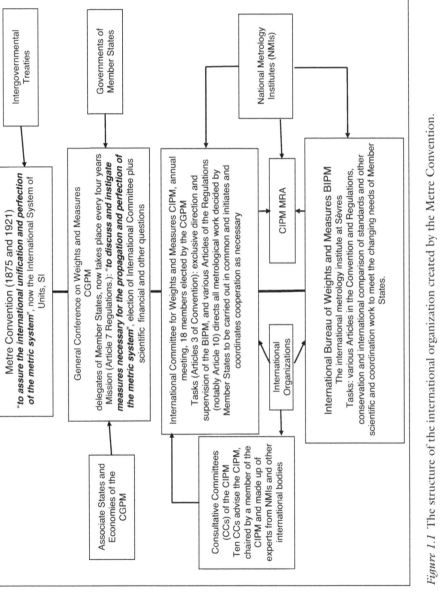

Figure 1.1 The structure of the international organization created by the Metre Convention.

re-elected at every CGPM. No name was given to the overall organization, as this was not the practice at the time. Beginning in the 1920s the CIPM began to create an expert Consultative Committee to advise it on the increasingly complex matters in the various areas of metrology; there are now ten of these. Further important developments in the organization of international metrology took place in the 1990s, but before coming to these we must first look at the creation and development of national laboratories for metrology, originally known as national standards laboratories.

In 1887, soon after the signing of the Convention, the German Government established the *Physicalisch-Technische Reichsanstalt* (PTR) in Berlin as the first national standards laboratory. This was followed in 1900 by the foundation of the National Physical Laboratory (NPL) in Teddington (UK) and the National Bureau of Standards (NBS) in Washington DC (USA) in 1901. The role of governments in metrology thereby became firmly established. Today all industrialized countries of the world, as well most developing countries, have a national institute charged with maintaining and disseminating national standards. These have become known as National Metrology Institutes, or NMIs.

The scientific nature of the work carried out in such an institute was clear from the beginning by the calibre of the staff and directors of these newly created institutes. The first President of the PTR was Helmholtz, and the Chairman of the Governing Board was Siemens. In Great Britain the first Director was Sir Tetley Glazebrook, Fellow of the Royal Society, and the Chairman of the Governing Board was Lord Rayleigh. In the USA it was the then-Secretary of the Treasury who took the initiative and the first Director of the new Bureau of Standards, later to become the National Bureau of Standards, NBS, and more recently National Institute of Standards and Technology, or NIST.

The range of metrological activity carried out under the auspices of the Metre Convention, i.e., the range of the international metric system, has greatly expanded since 1875. The first attempts to move beyond length, mass and temperature (already necessary for the thermal properties of the Metre) were not long in coming. In 1881, at the first international Congress of Electricians, which took place in Paris, it was proposed that the Metre Convention should take responsibility for electrical standards. Despite a very large majority of delegates in favour of the proposal, it met with opposition, essentially by those who only six years previously had objected to the creation of the BIPM as a scientific organization. The argument was that the BIPM would not be a suitable institution to take part in the very active scientific work then in progress aimed at establishing electrical standards. A formal proposal was made at a conference that took place in 1882 to define electrical units, but it was not accepted, the same people being present then as in 1881 and 1875.

During the first decade of the twentieth century, the British member of the CIPM, Sir David Gill, and the then-Director of the NBS, Samuel Stratton, both made great efforts to enlarge the role of the Metre Convention to include all areas of science where there was a need for international work in metrology. They did not succeed, and it was not until the 6th General Conference on Weights and

Measures in 1921 that agreement was reached to include electrical standards. There was still no formal name for the international metric system. Photometric standards soon followed in 1927, and in the 1960s ionizing radiation standards were added. It was only at the 11th CGPM in 1960 that, as well as redefining the Metre and the Second, the name *International System of Units* was given to the international metric system. Responsibility for time standards came to the BIPM only in 1987. Similarly, the mole, a unit of amount of substance, was defined by the CGPM in 1971, but the responsibility of the CIPM for metrology in chemistry started only in 1993.

Finally, at the 21st General Conference in 1999, the Member States accepted a proposal from the CIPM that the Metre Convention should have authority to take action in any field of science for which there was a need for international work in metrology. This proposal came in a wide-ranging report on future activities and needs for metrology drawn up by W. R. Blevin, at that time Secretary of the CIPM. As a consequence of this extension of the activities to be carried out under the Convention, the BIPM was instrumental in the creation of a Joint Committee for Traceability in Laboratory Medicine (JCTLM), which brought together the BIPM with the International Federation of Clinical Chemistry and Laboratory Medicine (IFCC) and the International Laboratory Accreditation Cooperation (ILAC), as well as representatives of manufacturers and regulators in the field of clinical chemistry. In order to formalize some of its relations with international bodies having links with metrology, the BIPM has signed memorandums of understanding with several worldwide organizations such as the International Organization for Legal Metrology (OIML), ILAC (International Laboratory Accreditation), the World Meteorological Organization (WMO), the International Atomic Energy Agency (IAEA) and the World Health Organization (WHO). In addition, it has active cooperation with several others, as a representative in working groups or commissions. These expanded areas of activity did not lead to any additional units of the SI because the set of seven base units was found sufficient to meet the metrological needs in all areas of science.

3   The evolving role of a National Metrology Institute

At the time of the 21st CGPM in 1999, the directors of 38 National Metrology Institutes and one international organization signed an agreement for mutual recognition of their national standards and calibration capabilities. This became known as the CIPM MRA (CIPM Mutual Recognition Arrangement) and fulfils an important role in international metrology, which we shall come to later. We must first, however, look at the changing role of National Metrology Institutes.

Originally, the role of the first such institutes, the PTR, NPL and NBS, were quite clear. It was to support national manufacturing industries, to establish national measurement standards, to provide calibrations and, where necessary, to ensure comparability with the national standards of other countries for the purposes of international trade – the most important of all these being the first. Indeed, both the NPL and the NBS were created in part because their governments had been persuaded that the success of the PTR was giving German industry an "unfair" advantage! In those days a clear hierarchical chain existed for almost all measurement standards, extending from the national standard to the workshop bench. Traceability, in the sense of a continuous chain of calibration certificates accompanying material artefacts, soon extended throughout individual nations and across the world through occasional international comparisons of national standards. In this the BIPM played a key role for length and mass, of course.

Up until about the 1970s, most high-level calibrations for industrial clients were carried out by the national laboratories themselves. This then became increasingly difficult as the number of such calibrations outstripped the resources of these laboratories. National calibration services were set up, comprising networks of industrial and commercial calibration laboratories that themselves were linked to the national laboratories. Today, very few industrial calibrations are carried out directly for industrial clients. The role of the National Metrology Institute in this respect is now to provide national standards and disseminate them through calibrations to the national calibration service. In so doing, however, the NMI carries out an additional and essential task, namely the dissemination of expertise in measurement and calibration. One of the ways this takes place is through the participation of experts from the NMI in the evaluation of the competence of the calibration laboratories. Thus, although the total number of calibration certificates issued by an NMI now is much lower than in the past, each certificate going to a calibration laboratory is the reference for hundreds if not thousands of calibration certificates issued by that calibration laboratory.

Traceability of measurement results means that a given result is obtained in terms of measurement units that are linked by an unbroken chain of calibrations or comparisons to national measurement standards – in practical terms – to SI units. At each link of the chain the uncertainty of the calibration or comparison must be given. In this way, a proper uncertainty of the final measurement in terms of SI units can be given. It is only by knowing the uncertainty associated with a measurement (commonly understood to signify the inverse of its accuracy: small uncertainty = high accuracy = high reliability) that the user can decide whether or not it is suitable for the application in hand. The traceability chain may be long, with many intervening calibrations through a complex hierarchy of standards, or it may be short, with just one calibration from the national metrology institute.

In some domains, such as in voltage or laser wavelength standards, it is now common for industrial users to have direct access to atomic or quantum-based standards of the highest accuracy. Such standards are sometimes called intrinsic standards. For the most demanding users, the former hierarchical system of standards is thus disappearing and being replaced by a system of comparisons, which verify that these independent commercial primary standards are operating correctly. This is an essential step because these so-called intrinsic standards do not *per se* guarantee accuracy, as there are innumerable errors that can be made in their construction and operation. The role of the national laboratory is, therefore, to provide the means of making these comparisons and to ensure that its own standards are closely linked with those of other countries, these days through the CIPM MRA.

Measurement standards are not static. They evolve continually to reflect advances in science and in response to changing industrial and other needs. It is necessary, therefore, for an NMI to maintain an active research base in measurement science. Research in measurement science is a long-term activity that must necessarily be done in advance of industrial and other requirements. It is said that today's research provides the base for tomorrow's calibration services.

The national benefits of an active research base in an NMI are not only long term, they are also immediately available through the expertise of the staff that comes only from being active in research. Major NMIs have thousands of industrial visitors each year; they run courses and seminars and are represented on all important industrial standards bodies. These close contacts with national industry also provide some of the essential knowledge for the NMI on present and future industrial requirements and the technology transfer to users in the country. The advantages to be gained by having all measurement standards in one institute are now widely recognized. A considerable synergy now exists between the many areas of metrology, and an institute that contains them all stands to gain significantly, not only in efficiency, but also in the quality and vibrancy of its science.

4   The CIPM MRA to meet the needs for confidence in worldwide uniformity of measurement

The international comparison of national measurement standards has always been an important activity of the national laboratories, but at the beginning of the 1990s it became clear that a more formal and structured approach was becoming necessary. This was in part a consequence of the widespread implementation of quality systems based on ISO standards, but also a result of the increasing demands for comparability and mutual recognition of measurements made in different parts of the world. The range and complexity of measurement requirements, already mentioned at the beginning of this chapter, is illustrated in a little more detail in the following list:

- national and international trade increasingly requires demonstrated conformity to written standards and specifications with mutual recognition of measurements and tests, i.e., worldwide traceability of measurement results to the SI;
- the economic success of most manufacturing industries is critically dependent on how well products are made, a requirement in which measurement plays a key role;
- navigation, telecommunications, now becoming an increasingly important part of today's world, require the most accurate time and frequency standards;
- human health and safety depend on reliable measurements in diagnosis and medical treatment, and in the production and trade in food and food products;

- the protection of the environment from the short-term and long-term destructive effects of industrial activity can only be assured on the basis of accurate and reliable measurements;
- global climate studies depend on reliable and consistent data from many disciplines over long periods of time, and this can only be assured on the basis of measurement standards linked to fundamental and atomic constants;
- physical theory, upon which all of this rests, is reliable only to the extent that its predictions can be verified quantitatively, and this calls for measurements of the highest accuracy;
- and finally, metrology has been shown to provide a high rate of return on investment.

The result of many discussions that took place in the early 1990 and following a succession of meetings of the Directors of National Metrology Institutes organized by the BIPM, starting in 1997, was the "Mutual Recognition Arrangement for national measurement standards and of calibration and measurement certificates of national metrology institutes",[3] the so-called CIPM MRA. This was signed on 14 October 1999 in Paris by the directors of 38 National Metrology Institutes and two international organizations. In summary its purpose and operation are as follows:

Stated objectives of the MRA are:

- to establish the degree of equivalence of national measurement standards maintained by the national metrology institutes (NMIs);
- to provide for mutual recognition of calibration and measurement certificates issued by NMIs;
- thereby to provide governments and other parties with a secure technical foundation for wider agreements related to international trade, commerce and regulatory affairs.

The process for meeting these objectives is as follows:

- international comparisons of measurements, to be known as key comparisons;
- supplementary comparisons of measurements;
- quality systems and demonstrations of competence by NMIs.

The planned outcome is:

- statements of the measurement capabilities of each NMI in a database maintained by the BIPM and publicly available on the web.

This can be summed up as a process to give mutual confidence between national metrology institutes as to the accuracy and reliability of their standards and calibration and measurement certificates. It is also a process that was designed to give the same confidence to the outside users of the services of the national laboratories. A key role in the operation of the MRA is played by the Regional Metrology Organizations. These began with collaboration between the NMIs in the Asia-Pacific region, the Americas and in Europe, but they now extend wider, as regional collaborations cover most regions of the world.

The CIPM MRA has transformed international metrology. The CIPM MRA in 2018 has been signed by the representatives of 59 Member States, 41 Associates of the CGPM and four international organization – and it covers a further 156 institutes designated by the signatory bodies.[4] The international organizations that are signatories are the International Atomic Energy Agency, the World Meteorological Organization, the European Space Agency and the European Joint Research Centre.

5   Accuracy rather than simply reproducibility or precision: the need to base units on fundamental constants

The question is often posed as to what extent it is necessary to talk about accurate or "absolute" measurement when surely all that is needed is some sort of comparability across the world, leading to reproducibility over reasonable periods of time. Surely this can be achieved without any call upon measurements linked to atomic physics or fundamental constants. In other words, why is metrology today so complex and expensive? Has it not all already been done?

What is not understood by those putting forward these sorts of arguments is that it is only by means of accurate measurements, by which we mean those that provide a close representation of nature, that the apparently simple requirement for comparability and long-term reproducibility can be met. Accurate measurements are those made in terms of units firmly linked to fundamental physics so that they are (a) repeatable in the long term and (b) consistent with measurements made in other areas of science and technology. They are thus much more than merely reproducible or precise, which are measures of how well successive measurements agree with each other. One can have an instrument that gives very reproducible or precise measurements but are offset from the true value by a calibration or other instrumental error. In this case the results are precise but not accurate. If they are offset by such an unknown calibration or instrumental error, it is evident that the resulting measurements are not accurate and therefore not reliable.

Measurement standards based upon material artefacts cannot provide the assurance of long-term stability because material artefacts by their very nature are not immutable. Indeed, the present principal weakness of the SI in this respect is our inability to establish the long-term stability of the kilogram, and in consequence the multitude of other units that are derived from it such as mechanical, electrical and optical power and energy, to give just a few. The precision with which measurements are made depends on the application. The accuracy and the precision must be matched. It serves no useful purpose if the precision is not sufficient even if the accuracy is high, but a precise measurement can be confusing if the accuracy does not match the precision.

The problem of obtaining measurement standards that are stable in the long term was set out more than 100 years ago by Maxwell during his presidential address to the Physics Section of the British Association for the Advancement of Science in 1870:

> Yet, after all, the dimensions of our earth and its time of rotation, though, relatively to our present means of comparison, very permanent, are not so by physical necessity. The Earth might contract by cooling, or it might be enlarged by a layer of meteorites falling on it, or its rate of revolution might slowly slacken, and yet it would continue to be as much a planet as before. But a molecule, say of hydrogen, if either its mass or its time of vibration were to be altered in the least, would no longer be a molecule of hydrogen.
>
> If, then we wish to obtain standards of length, time, and mass which shall be absolutely permanent, we must seek them not in the dimensions, or the motion, or the mass of our planet, but in the wavelength, the period of vibration, and the absolute mass of these imperishable and unalterable and perfectly similar molecules.[5]

At the time neither science nor technology were adequate to meet Maxwell's precept. The establishment of accurate and practical measurement standards linked to fundamental constants, having also the range and diversity required for the whole of modern science and technology, is a major undertaking. Many areas of advanced metrology are now linked directly or indirectly to fundamental or atomic constants using techniques at the frontiers of science; we have atomic clocks using trapped ions or cold atoms or Bose-Einstein condensates, laser wavelength standards, femtosecond spectroscopy, quantum electrical standards, isotope dilution mass spectrometry, ultraviolet spectroscopy, nanometrology, atomic interferometry and many others, all of which require highly trained physicists. The days when standards of length, mass, time and electricity were totally separate and dependent on quite different technologies are now past.

6   The 2018 redefinition of the SI

The International System of Units, formally adopted by the 11th CGPM in 1960, was the culmination of more than a century of study and discussion on how best to establish a system of units that would bring together mechanical and electrical units. Today, the SI includes the seven base units, the second, metre, kilogram, ampere, kelvin, mole and candela as well as derived units made up of algebraic combinations of the base units, multiples and submultiples and rules for their use. All this is laid out in a document approved by the CIPM and published by the BIPM under the title of *The International System of Units SI*. It is generally referred to as the *SI Brochure*. The *Brochure*, a document of some 75 pages, is now in its 9th edition (2019), and it gives a complete description of the SI. The full text in French and in English is freely available on the BIPM website, and a brief history of the development of ideas during the nineteenth and early twentieth centuries related to units is given in the Introduction.

In brief, while there were many advances in definitions of the seven base units of the SI, neither the idea of the great savants of the eighteenth century nor the precepts of Maxwell could be fully realized until now. This was mainly because there had been no way of defining a unit of mass other than as the mass of a particular artefact. Until the redefinition in 2018, the kilogram was still defined as the mass of the International Prototype of the Kilogram kept at the International Bureau of Weights and Measures at Sèvres.

However, this problem has now been solved and a fundamental revision of the SI has taken place. The 26th CGPM in 2018 adopted a CIPM proposal to redefine the SI in terms of a set of seven defining constants of nature each having an exact defined numerical value. Taken together they define the same seven base units of the SI that we had before, but they no longer include any material artefacts. This is laid out in Resolution 1 of the 26th CGPM entitled:

On the revision of the International System of Units (SI)

The General Conference on Weights and Measures (CGPM), at its 26th meeting, considering: (here there are a number of short paragraphs stating the needs and advantages of the new definition, not reproduced here) decides that, effective from 20 May 2019, the International System of Units, the SI, is the system of units in which:

- the unperturbed ground state hyperfine transition frequency of the caesium 133 atom $\Delta \nu_{Cs}$ is 9,192,631,770 Hz,
- the speed of light in vacuum c is 299,792,458 m/s,
- the Planck constant h is $6.626\,070\,15 \times 10^{-34}$ J s,
- the elementary charge e is $1.602\,176\,634 \times 10^{-19}$ C,
- the Boltzmann constant k is $1.380\,649 \times 10^{-23}$ J/K,
- the Avogadro constant N_A is $6.022\,140\,76 \times 10^{23}$ mol^{-1},
- the luminous efficacy of monochromatic radiation of frequency 540×1012 Hz, K_{cd}, is 683 lm/W,

where the hertz, joule, coulomb, lumen, and watt, with unit symbols Hz, J, C, lm, and W, respectively, are related to the units second, metre, kilogram, ampere, kelvin, mole, and candela, with unit symbols s, m, kg, A, K, mol, and cd, respectively, according to Hz = s^{-1}, J = m^2 kg s^{-2}, C = A s, lm = cd m^2 m^{-2} = cd sr, and W = m^2 kg s^{-3}.

In the list given here the exact numerical values in Resolution 1 are taken from the CODATA 2017 special evaluation of the constants made for this purpose. The definitions of each of the seven base units of the SI can be deduced from the ensemble of these constants. Before coming to this, there are two points to be addressed: what has changed that allow us to make this big change, and what does it mean to define an exact numerical value of a fundamental constant?

First, what changed that allowed us to make this big change in the definitions of our units, particularly in the definition of the unit of mass, the kilogram? The problem had always been that while two mass standards are easily compared to high accuracy with a simple beam balance, there had been no way to compare the gravitational force acting on a mass with any constant of nature. The key advance in science that opened the way to do this was the discovery of the quantum Hall effect by Klaus von Klitzing in 1980, which allowed an electrical resistance to be made whose value was exactly proportional to h/e^2. Taken together with the Josephson effect, discovered in 1962, giving a voltage exactly proportional to the ratio $h/2e$, it became possible to establish values of electrical voltage and resistance at the levels of millivolts and thousands of ohms, and hence an electrical current, related to quantum effects. The ratio $2e/h$ was designated the Josephson constant K_J and the ratio h/e^2 was designated the von Klitzing constant R_K. It thus became possible to produce not only an electric current but also other electrical quantities, notably power, proportional to a combination of fundamental constants at a level where they produced visible and accurately measurable physical effects in the laboratory. These are known as macroscopic quantum effects because the electrical resistance and voltages produced are indeed macroscopic and useable in technology.

The device that opened the way to a quantum definition of the unit of mass based on these two macroscopic quantum effects had been invented in 1974 by Bryan Kibble of the National Physical Laboratory (NPL) in the UK. He had a clever idea on how to use a simple equal arm balance to compare electrical and mechanical power to high accuracy. It was very simple in principle and is now used to balance the gravitational force acting on an object, such as a 1 kg mass standard, against an electromagnetic force developed by an electrical current passing through a coil in a magnetic field. The electrical current is measured in terms of a voltage and resistance linked to the Josephson and quantum Hall effects, the measured quantities being, mass, the acceleration due to gravity, the velocity of a moving coil and microwave frequencies needed to make the Josephson effect work. The end result is an equation linking mass, velocity, the acceleration due to gravity, a microwave frequency and the Planck constant, as we shall see. One can say that the two macroscopic quantum effects make, in one step, the enormous jump in scale between the microscopic world of quantum phenomena and the macroscopic world of kilogram mass standards. The whole apparatus, originally known as a watt balance, is now referred to as a Kibble balance since Bryan's untimely death in 2016.

But this was not all; in a totally different domain of science, advances were made that opened the way to linking the kilogram to the mass of an atom. In the 1970s, Deslattes at the National Bureau of Standards, and Hart and Bonse in the UK and Germany, made direct measurements of the lattice spacing of silicon by combining x-ray and optical interferometry. This key advance made it possible to conceive of an experiment in which the number of atoms in a sample of silicon could be determined to high accuracy by weighing an artefact whose volume had been determined. This leads to a value for the Avogadro constant, the number of atoms in 1 mole of silicon. The measured quantities are the lattice spacing of the atoms in a crystal of silicon, the mass and volume of the artefact and the molar mass of silicon. Silicon was the only element that industrial technology had been developed to produce kilogram sized monocrystal samples. Since we know the relative atomic mass of silicon to that of carbon 12, this became a second, quite independent, path to defining a unit of mass in terms of a fundamental constant, namely the Avogadro constant and hence mass of an atom of silicon. This is known as the x-ray crystal density method. Through an equation of physics, which we shall come to later, the Planck constant and the Avogadro constant are linked so that the results of Kibble balance experiments and x-ray crystal density experiments can be directly compared.

Despite the simplicity of the principles of these two routes to a new definition of the kilogram, it took nearly 30 years to reach the stage at which the results of these experiments had accuracies and consistencies sufficient enough to draw up a detailed proposal for the new, absolute system of units.

The implementation of the Josephson and quantum Hall effects in the 1980s and 1990s had, however, another and much more immediate consequence. It very quickly became possible to maintain reference standards for the volt and the ohm, based on these two effects, whose reproducibility approached parts in 10^{11}. Their absolute values in terms of the SI volt and ohm were, however, known only to a few parts in 10^7, limited mainly by uncertainties in the value of the Planck constant. These quantum-based reference standards were so useful that in 1990 the CIPM adopted conventional values for K_J and R_K that were designated K_{J-90} and R_{K-90} respectively. By so doing, however, electrical metrology became much more precise but became, in one sense, decoupled from the other units of the SI at the highest levels of accuracy.

The second point to be addressed is the following: how is it possible to arbitrarily define an exact numerical value of a fundamental constant when, by its very nature, its value is set by nature and not by us? The important distinction to be made is the following, the value of a constant is indeed fixed by nature, but its numerical value depends on the size of the unit in which we choose to measure it. Let us take the example of the speed of light, c, which is indeed a constant of nature and is the same everywhere, but its numerical value depends on the size of our units of time and length. For example:

c = 299 792 458 metres per second
or c = 983 571 056.4 feet per second
or c = 327 857 018.8 yards per second

the value of c = numerical value × unit

Thus, if we define the units metre and second independently, then we must determine the numerical value of c by experiment, and it will have an uncertainty. That was the situation before 1983, when both the metre and the second were independently defined.

But, if the second is independently defined in terms of the frequency of the caesium transition, and we choose to define the numerical value of c, to be equal to 299 792 458 exactly, then the effect is to define the size of the remaining unit, the metre. This is the current definition of the metre, since the change in 1983. The numerical value of the speed of light in metres per second now has zero uncertainty.

The more complicated case is that of the kilogram, for which we decided to take the Planck constant, h, as the reference:

$h = 6.626\ 070\ 15 \times 10^{-34}$ kg m² s⁻¹

the value of h = numerical value × unit

Thus, if we define each of the units kilogram, metre and second independently, then we must determine the numerical value of h by experiment, and it will have an uncertainty. This was the situation before 2018. The numerical value was determined by a combination of Kibble balance and silicon density experiments.

But, since the metre and the second were already independently defined, we could choose to define an exact numerical value of h, then the effect was to define the kilogram. This became the proposed new definition of the kilogram. We just had to make sure that we chose the right value of h, i.e., one that was really consistent with the then definition of the kilogram, which is why it was done in two independent ways through the Kibble balance and silicon crystal density experiments.

All this was already proposed in detail to the 24th CGPM in 2011 and after much discussion, prior to the Conference in the CIPM Consultative Committee for Units (CCU), the proposal received unanimous approval by Member States. A certain number of conditions were set before such a scheme could be formally adopted. Principal among these was that consistent data be obtained that would justify confidence in the numerical values to be adopted for the constants in question. Consistency among the values of constants is in any case needed for science and technology in general. This has, since the 1970s, been overseen by the Task Group on Fundamental Constants group of CODATA, the Committee on Data for Science and Technology, which operates under the auspices of the International Council for Science, or ICSU. Periodically, now every four years, the Task Group publishes recommended values for the fundamental constants. The most recent is dated 2017 and, pending publication in the *Reviews of Modern Physics*, can be found on the NIST website at physics. nist.gov/constants.

The key point of consistency required before one could be confident that the numerical value of the Planck constant could be accepted for the New SI was that the values of h obtained via the Kibble balance and the silicon crystal density be consistent at the level of a few parts in 10^8. The measurements of the fundamental constants that go into the CODATA list are all measured with respect to the present base units of the SI. The relationships between fundamental constants are through the equations of physics, which are independent of the units with which they are measured. Thus, the test of the consistency that we required was through an equation that links the Planck and Avogadro constants, which included other constants of physics provided that their values were independently known to high accuracy. Thus a comparison of the value of h given by the Kibble balance and that deduced from measurements of the Avogadro constant provided not only a critical test of the accuracy of the two measurements but also of the consistency of all the physics involved.

In the Kibble balance the comparison of electrical and mechanical power leads to a simple equation $mgv = IU$ where m is a mass of about 1 kg, g is the acceleration due to gravity, v is the measured velocity of a coil, I is the current needed to balance the gravitational force of the mass on the pan and U is the voltage produced by the coil moving through a magnetic field. These are all simple macroscopic electromechanical quantities, but by means of the Josephson effects the quantity U can be represented by $nfh/2e$ where n is an integer and f the frequency of the microwave electromagnetic radiation used to produce the voltage in the Josephson junction. Using also the quantum Hall effect, which gives an electrical resistance $R = h/ie^2$, we can write

$$IU = in^2f^2h/4 \text{ so that } mgv = in^2f^2h/4 \qquad (1)$$

or

$$h = 4mgv/in^2f^2. \qquad (2)$$

For the silicon crystal density, we have:

$$N_A = n\, M(\mathrm{Si})/\rho\, a^3 \qquad (3)$$

where here n is the number of atoms per unit cell of silicon, $M(\mathrm{Si})$ the molar mass of silicon, ρ the density of the sample of silicon and a its lattice constant.

The equation linking the Avogadro constant N_A to h is the following:

$$N_A h = [c\alpha^2/2R_\infty][M_u\, A_r(e)], \qquad (4)$$

where α, R_∞, M_u and $A_r(e)$ are, respectively, the fine structure constant (known to parts in 10^{10}), the Rydberg constant for infinite mass (parts in 10^{12}), the molar mass constant (exact) and the relative atomic mass of the electron (parts in 10^{10}).

We can thus write for the silicon density experiment, the value of h deduced from the silicon density measurement of the Avogadro constant is

$$h_{(\mathrm{silicon})} = [c\alpha^2/2R_\infty][M_u\, A_r(e)]\, \rho\, a^3 / n\, M(\mathrm{Si}) \qquad (5)$$

This must now be compared with the Kibble balance result:

$$h_{(\mathrm{Kibble\ balance})} = 4mgv/in^2f^2 \qquad (6)$$

The important question was: how well do these two methods of arriving at a value for h agree? The answer finally achieved in 2017 was that they agree within their respective uncertainties, namely, a few parts in 10^8.

What does this demonstrate? The most important outcome is the demonstration that the Josephson and quantum Hall relations correctly represent macroscopic voltages and resistances – something that had not been demonstrated at this level before. It also demonstrates a remarkable level of consistency among the measured values of fundamental constants using a wide variety of methods based on an equally wide variety of equations of physics. One can conclude that classical and quantum physics in these areas are consistent to a few parts in 10^8.

On a much more practical level, this agreement opened the way to redefine the kilogram. By fixing the numerical value of h, the kilogram can be realized by methods based on either of these two equations with an accuracy sufficient for all practical needs, namely a few parts in 10^8. This result is the key to the New SI and was finally achieved and agreed by all at the meeting of the Consultative Committee for Units in September 2017.

7   The practical realization of SI units

An essential prerequisite for any system of units is that it be practical, i.e., it can be used to make measurements at any level of precision or accuracy required. The experimental methods used for the realization of units using the equations of physics as illustrated earlier are known as primary methods. The essential characteristic of a primary method is that it allows a quantity to be measured directly from the definition of its unit without prior knowledge of that unit by using only constants that themselves do not contain that unit. The Kibble balance and silicon crystal density methods are examples of primary method for the measurement of mass. In the definition of the SI, methods for doing this are much more general and open the way to any imaginable improvements in accuracy, which will not be limited by the definitions of the units, this was not the case in the past.

7.1   *In the SI before 2018*

In the past, a unit for a given quantity was taken to be a particular example of that quantity, chosen to be of a convenient size. Until very recently in human history, units were necessarily defined in terms of material artefacts, notably the metre and kilogram for length and mass, or the property of a particular object, namely the rotation of the Earth for the second. It was only in 1960 that the first non-material definition was adopted, namely the wavelength of a specified optical radiation for the metre. Until 2018, definitions of the ampere, kelvin, mole and candela had been adopted that no longer referred to material artefacts but, in the case of the ampere, to a specified electric current required to produce a given electromagnetic force (although the unit of force was linked to the kilogram) and for the kelvin to a particular thermodynamic state, the triple point of water.

Even the atomic definition of the second was in terms of a specified transition of the atom of caesium. The kilogram had always stood out as the one unit that had resisted the transformation from an artefact. The definition that opened the way to real universality was that of the metre in 1983 which implied, although it did not state, a fixed numerical value for the speed of light. The definition was worded, however, in the traditional form and stated essentially that the metre was the distance travelled by light in a specified time. In this way it reflected the other definitions of the base units of the SI each of which had the same form, such as "the ampere is the current which", "the kelvin is a fraction of a specified temperature" and so on. Such definitions can be called explicit unit definitions. Although they met many of the requirements for universality and accessibility, and a variety of realizations are often possible, they nevertheless constrained practical realizations to experiments directly or indirectly linked to the particular conditions or states specified in the definitions. In consequence, the accuracy of realization of such definitions could never be better than the accuracy of realization of the particular conditions or states specified in the definitions.

7.2   In the current SI since 2018

The practical realization and the making of measurements using the present, revised, SI are in principle quite different. Instead of each definition specifying a particular condition or state, which as we have seen sets a fundamental limit to the accuracy of realization, it is now open to us to choose any convenient equation of physics that links the particular constant or constants to the quantity we want to measure. This is a much more general way of defining the basic units of measurement. It is one that is not limited by today's science or technology, as future developments may lead to at present unknown equations that could result in quite different ways of realizing units to much higher accuracy. The exception remains the definition of the second in which the original microwave transition of caesium remains, for the time being, the basis of the definition.

For the kilogram, the unit whose definition has undergone the most fundamental change, realization can be through any equation of physics that links mass, the Planck constant, the velocity of light and the caesium frequency. These have been illustrated already in the previous section with respect to the Kibble balance and the silicon density method. Another possibility for measuring mass through the new definition, but this time at the microscopic level, is through measurements of atomic recoil using the relation that includes h/m.

A question that might be asked is what are the consequences for the National Metrology Institutes now that the base units of the SI, referred to constants of nature, can in principle be realized by anyone anywhere independently of their local NMI? The answer is that the effect is very small, as it has been for independent realizations of the Josephson and quantum Hall devices. First, the effort and expertise needed to realize a base unit of the SI to a useful accuracy is itself quite large and few people or institutions are capable of it. But even if they can, for their realization to be recognized by others it must be linked to the world measurement system through participation in the CIPM MRA. Any independent realization of a unit, even based on one of the macroscopic quantum effects, can have validity only by comparison with one that has been linked to the world network. The possibilities for error in setting up a Josephson or quantum Hall system, or an atomic clock for that matter, are such that without independent verification it can have very little value. Such systems are always linked through comparisons with national standards. They are not "calibrated" in the normal sense because they are not given a specific value other than the intrinsic one, but an estimate is given of the accuracy with which they reproduce the intrinsic value of a Josephson system. This is now the situation for standards derived from the new definitions of the base units of the SI.

In the SI today, there are no limits to the methods that human ingenuity can devise for realization of its units. James Clerk Maxwell would be pleased with this!

8   Conclusions

Since the creation of the metric system and the beginning of mass production of engineering products, metrology has developed to become a key component of the technical infrastructure of the modern world. The industrialized nations of the world have put in place and support a worldwide network of laboratories that together provide the technical basis for a multitude of the essential constituents of everyday life.

The way in which the national measurement infrastructure is organized and how it is financed are, of course, matters for individual governments to decide. What is sure, however, is that an advanced industrial economy must have access to measurement standards: the government and industry must have access to advice on measurement matters; there must be experts qualified to represent national interests on international bodies concerned with measurement; and there must exist the research base in measurement science, without which none of this is possible. Participation in international activities, notably the CIPM MRA and all that goes with it, are now essential.

At the end of the nineteenth century, far-sighted men clearly understood the link between the economic success of manufacturing industry and access to accurate measurement standards, and the need for research to allow these standards to advance. Since then, the accuracies required, and the range of applications requiring accurate measurement, have increased almost beyond recognition, but the basic arguments for a national measurement infrastructure remain today exactly as set out at the end of the nineteenth and beginning of the twentieth centuries.

Notes

1 This chapter is a revised and extended version of *The Development of Modern Metrology and its Role Today* by T. J. Quinn and J. Kovalevsky, which appeared in *Philosophical Transactions of the Royal Society* Vol 363 No 1834, pp 2307–2327, (2005), the proceedings of a discussion meeting held at the Royal Society on 14 and 15 February 2005 entitled *Fundamental constants of physics, precision measurements and the base units of the SI*.

2 Documents Diplomatique de La Conférence du Mètre (Paris, Imprimerie Nationale 1875).
3 Resolution 2 of the 21st CGPM, BIPM, 1999.
4 See the BIPM website: www.bipm.org for up to date information on the CIPM MRA.
5 Maxwell, J. C. (1870). British Association for the Advancement of Science, Liverpool, 1870, Transactions of the Sections, Notices and Abstracts of Miscellaneous Communications to the Sections, Mathematics and Physics.

References

The BIPM website – www.bipm.org – includes a great deal of information on the Metre Convention, General Conferences on Weights and Measures, the International Committee for Weights and Measures, its Consultative Committees, the CIPM MRA, the SI and the BIPM. On this same website there are links to National Metrology Institutes and Regional Metrology Organizations.

For a detailed history of the origins of the Metre Convention and the BIPM see also Quinn, 2011. Three recent papers that deal in detail with the Metre Convention and with the development of the revised SI are Quinn (2017a, 2017b, 2017c). For the 2017 CODATA recommended values of the fundamental constants of physics see: http://physics.nist.gov/cuu/Constants.

Quinn, T. J. (2011). *From artefacts to atoms, the BIPM and the search for ultimate measurement standard.* New York, NY: Oxford University Press.

Quinn, T. J. (2017a). The Metre Convention and the BIPM. In P. Tavella, M. J. T. Milton, M. Inguscio, & N. De Leo (Eds.), *Proceedings of the International School of Physics 'Enrico Fermi', Course 196: Metrology: From physics fundamentals to quality of life.* Amsterdam: IOS; Bologna: SIF.

Quinn, T. J. (2017b). The development of units of measurement from the origin of the metric system in the 18th century to proposals for redefinition of the SI in 2018. In P. Tavella, M. J. T. Milton, M. Inguscio, & N. De Leo (Eds.), *Proceedings of the International School of Physics 'Enrico Fermi', Course 196: Metrology: From physics fundamentals to quality of life.* Amsterdam: IOS; Bologna: SIF.

Quinn, T. J. From artefacts to atoms – A new SI for 2018 to be based on fundamental constants, *Studies in History and Philosophy of Science, 65–66,* (2017), 8–20.

2 Justifying and motivating an SI for all people for all time

Martin J. T. Milton

1 Introduction – metrology and the role of the SI

The global aim of those working in metrology is stated in the vision statement of the *Bureau International des Poids et Mesures* (BIPM) as being "to promote and advance the global comparability of measurements". One of the principal ways this has been achieved is by the agreement of a system of units for use in all applications of measurement around the world. The foundation of the system is a set of base units each of which has a definition and an agreed protocol as to how it should be put into practice (known as a *mise en pratique*). These base units, together with a set of units derived from them, and a set of prefixes, form what is called the International System (or SI).

Although the aim of metrology is to provide the basis for measurements that are unchanging over long periods of time, it has always been a practical and dynamic enterprise that has exploited scientific advances in order to improve measurement results and to increase the effectiveness with which accurate measurements can be disseminated (Milton, Williams, & Bennett, 2007). Recent manifestations of the desire to improve the global basis for stable and comparable measurements are the proposals to revise the definitions of some of the base units of the SI. They are often referred to as being the proposals for a "New SI" and are the subject of this chapter.

2 The development of the SI

The title International System (SI) was officially given to the system of units developed by the BIPM in 1960. It is a system that has been changed whenever there has been consensus about the need for improvements. In particular, there have been three significant changes to the set of base units since 1960 that are relevant to the discussion presented here:

- In 1971, the mole was added to the system as the base unit for the quantity amount of substance. Thereby introducing a seventh base unit.
- In 1983, the definition of the metre was changed to be the distance travelled by light in a specified fraction of a second thus fixing the speed of light. This was the first example of a definition of a base unit that referred to a fixed numerical value for a specified fundamental constant.

- In 1990, a set of units was adopted for practical electrical measurements based on agreed ("conventional") values for the Josephson constant (K_J) and the von Klitzing constant (R_K). Their use superseded the use of the ampere for practical electrical measurements.

Whilst the first two of these had no significant drawbacks, the change made to the electrical units by introducing the 1990 "convention" had a significant drawback. This is because a central principle of the SI has always been that it is a coherent system of units. This means that measurement results expressed in terms of the seven base units, and the derived units can be combined consistently using the equations of physics and chemistry without the need for any additional conversion or scaling factors. However, the 1990 convention is not strictly coherent with the other SI units because the values of K_J and R_K as calculated from the best available values of e and h have diverged from the conventional values agreed to in 1990. This drawback with the 1990 convention proves to be a major argument in favour of finding a new definition for the ampere that would have the precision and stability brought by the 1990 convention, but without the implicit loss of coherence. This chapter concerns the proposals made in 2006 to adopt new definitions for some of the base units of the SI (Mills, Mohr, Quinn, Taylor, & Williams, 2006). Since that first publication, the proposals have been refined in several other publications and have gained very widespread support. They are based on a carefully argued set of technical advantages that have been debated intensely not least because they are accompanied by several technical disadvantages. The balance between these advantages and disadvantages is seen differently by users working in different fields. It also depends to some extent on the results of an extensive programme of laboratory experimentation at several National Metrology Institutes (NMIs) around the world. Many publications show that the results of this programme of experimentation either met, or were very close to meeting, the criteria laid down for the adoption of the new definitions (Milton, Davis, & Fletcher, 2014). Hence, the proposed changes were endorsed by the Conference Générale des Poids et Mesures (CGPM) on 16 November 2018.

In this chapter we describe the 2006 proposals and their technical consequences. We identify the motivations articulated for adopting them and consider how they have become sufficiently compelling for a consensus to be established around their adoption in just 11 years. In particular, we consider whether the arguments have largely been presented pragmatically (thus reflecting a balance of practical advantage versus disadvantage) or axiomatically (by reference to externally established principles).

3   The 2006 proposals deconstructed – their technical advantages and disadvantages

This chapter concerns the proposals to redefine certain base units of the SI that were first published in detail by Mills et al. in 2006. The origin of these proposals is found in a paper published by the same authors the previous year under the title "Redefinition of the kilogram: a decision whose time has come" (Mills, Mohr,

Quinn, Taylor, & Williams, 2005) – a title that makes a specific reference to the historic publication on metrication, *A Metric America: A Decision Whose Time Has Come* (NBS, 1971), thus suggesting that the authors considered that they were proposing a matter of some historical significance.

The 2005 publication itself referred to a previous publication by two of its authors (Taylor & Mohr, 1999), which had recommended that the kilogram be redefined in terms of a fixed value of the Planck constant (h) justified at that time because of "the superiority of the watt balance to other methods of linking an invariant of nature to the kilogram". The 2005 publication broadened the 1999 proposal by arguing that the kilogram could be redefined in terms of a fixed value of either the Planck constant (h) or, alternatively, the Avogadro constant (N_A). Recognizing that the best values available at that time for these constants were inconsistent and insufficiently accurate to be the basis of a new definition, they proposed that such a definition might be facilitated by "adopting a conventional value for the mass of the international prototype, $m(K)_{07} = 1$ kg exactly, it could remain the basis for the worldwide system of practical mass measurement" (Mills et al., 2005).

The 2005 proposals are not the same as those made in 2006. They were motivated primarily by the desire to redefine the kilogram in order to avoid the influence of any possible unsatisfactory properties of the International Prototype of the Kilogram (IPK) on the accuracy with which the definition of the kilogram could be realized. These were widely thought to be the cause of changes in the differences between the mass of the IPK and the masses of the copies maintained with it (the *témoins*). When measured in 1989, these had changed by as much as 50 µg from the values of the same differences measured 100 years previously. By assuming these differences had accumulated linearly over 100 years, they have been equated to a drift in the mass of the IPK with an upper limit of 0.5 µg/year. Since these estimates presented a possible level of instability in the IPK that was less than the inconsistency between the best values of h and N_A, the authors of the 2005 publication proposed that a conventional value be used for the IPK. This would allow its value to be defined to be constant whilst the unit of mass was realized in practice through experiments capable of measuring either h or N_A.

The authors of the 2005 publication also argued that the adoption of their proposals would have the benefit of reducing the uncertainty of some of the fundamental constants. Even in 2005, the argument for the change was made very strongly by the authors stating their views that "the advantages of redefining the kilogram immediately outweigh any apparent disadvantages" (Mills et al., 2005). It was even suggested that it could be implemented at the next meeting of the CGPM in 2007. Later in 2005, the proposals were endorsed enthusiastically by the Consultative Committee on Units (CCU), which recalled the following in a recommendation:

> Consensus now exists on the desirability of finding ways of defining all of the base units of the SI in terms of fundamental physical constants so that they are universal, permanent and invariant in time.
>
> (CCU Recommendation No 1 2005)

Later the same year, the proposals were endorsed by the Comité International des Poids et Mesures (CIPM), which, in its Recommendation 1 (CI-2005), approved "in principle" the preparation of new definitions for the kilogram, ampere, kelvin and mole, although it gave no indication of the motivation for such a change, except indirectly through its references to the published recommendations of the CCU.

Despite the support for the proposals from the CCU and the CIPM, the consultative committees responsible for measurements of mass and temperature (the CCM and the CCT) feared a break in the continuity of measurements in their areas if the proposals were adopted too soon. Hence, they developed criteria that experimental data for h and k (the Boltzmann constant) should meet before they would be willing to agree that the proposals could be adopted (CCM, 2005; Gläser et al., 2010).

The following year the proposals were presented in full for the first time in a publication (Mills et al., 2006) that also included the authors' responses to criticisms made of their 2005 publication. It is this 2006 publication that is the subject of the discussion here. It updated the 2005 proposals by confirming the preference of the authors for the kilogram to be defined by fixing the numerical value of h, rather than N_A, and it added a proposal to reintroduce the electrical units into the SI. Of particular importance to the discussion presented here is that the authors stated that their detailed proposals were structured according to three "guiding assumptions". These were:

i   that the overall structure of the SI should be preserved, because "these quantities and units are deemed to meet current and future needs",

ii   that "it is not always necessary that a new definition of an SI base unit should allow the unit to be realized with a reduced uncertainty", and

iii   that "the units to be redefined and the constants to which they are to be linked should be chosen in such a way as to maximize the benefits to both metrology and science".

These guiding assumptions provide insight into how the authors balanced the factors motivating their proposed changes. Whilst the first and third of these assumptions are articulated in a way that supports the interests of a broad range of users of the SI, the second is not as easy to justify. The rationale provided by the authors for this read:

> The benefits to both metrology and science of replacing the current definition of the kilogram by one that links it to an exact value of the Planck constant h, and the current definition of the kelvin by one that links it to an exact value of the Boltzmann constant k, are viewed as far outweighing any marginal increase in the uncertainty of the realization of the SI unit of mass or thermodynamic temperature that might result.
>
> (Mills et al., 2006)

According to this reasoning, the benefits of linking the definitions of certain base units to specified constants are more important than the possible consequences of an increased uncertainty with which the units might be realized in practice. Although not stated, it might be presumed that the authors anticipated an increase in uncertainty in the short term, but that the route to a reduction in uncertainty in the longer term could be foreseen, perhaps through technical advances. Despite this possible unrecorded caveat, the published text raises the question as to whether the motivation for the proposals was purely pragmatic or at least in some part axiomatic. Was it proposed because it was expected to lead to improved measurements, or because linking definitions to the exact values of fundamental constants would represent progress towards some overarching axiom? We return to this question later.

It is now useful to summarize the technical content of the 2006 proposals. They advocated two major and interconnected changes. The first was to make the system of practical electrical measurements that are currently based on the 1990 convention coherent with the SI by replacing the conventional values for the Josephson constant (K_J) and the von Klitzing constant (R_K) with values based on fixed values of h and e. These values chosen for h and e would be the best available values from a least-squares fitting of the fundamental constants carried out immediately prior to the time of their adoption.

The second major change arose from the first. It is that experiments that were previously used to determine h in terms of standards of mass (such as by the Kibble balance or using the silicon XRCD method) would be constrained to work in the opposite mode. This would allow them to determine standards of mass, and even to realize the kilogram, in terms of an agreed fixed value of h. This change would only be acceptable if it could be done with a practically useful uncertainty. (It would also be possible to realize atomic-scale masses with respect to a fixed value of h using atomic recoil methods, but these could not be used to realize masses at a macroscopic level.)

These changes relating to the definitions of the kilogram and the ampere were combined in the 2006 paper with proposals for the five other base units; they were:

i   to redefine the kelvin such that primary thermometric methods could be used to realize the kelvin by reference to a fixed value of the Boltzmann constant;

ii to redefine the mole as the amount of substance of a fixed number of entities in place of the present definition which refers to the amount of substance of a specified mass of a pure isotope of carbon (Taylor, 2009; Milton & Mills, 2009);

iii to re-word the definitions of the second, metre and candela to align them with the explicit constant style. There would not be any change in the principle behind these definitions; they would simply be re-worded.

The authors also identified that their proposals might bring some technical disadvantages, of which the most prominent would be:

i   An extremely small uncertainty component would have to be added to the mass of the IPK when expressed in terms of the redefined kilogram. This would be propagated through to the uncertainty of all mass values that continued to be traceable to the IPK.

ii   There would be a small shift in the ohm and a larger shift in the volt when moving from the 1990 convention to the proposed new values (Fletcher, Rietveld, Olthoff, Budovsky, & Milton, 2014).

iii An extremely small additional uncertainty would be added to the molar mass constant (when expressed in kg/mol) and to the temperature of the triple point of water (when expressed in K).

iv   The change to the new definition of the mole (Milton & Mills, 2009; Milton, 2013) might be considered by some authors to change the rationale for retaining it as a base unit.

The proposed changes would bring immediate benefits for those concerned with practical electrical measurements, but the impact on practical mass measurements would not be as straightforward or as beneficial. Following the publication of the proposals in 2006, the Consultative Committee Concerned with Measurements of Mass (the CCM) again considered the impact of the proposed changes by reviewing progress against the criteria it had agreed upon in 2005. Their conclusion was subsequently summarized as follows:

> Although these conditions will lead to an increase in the best possible uncertainty attributed to a national platinum iridium standard . . . the CCM considers this acceptable as an upper limit in view of current practical uncertainty requirements and the possible instability of the IPK.
>
> (Gläser et al., 2010)

4   The proposed definitions

In this chapter, we refer prosaically to "the 2006 proposals". The term "New SI" was used in papers submitted to meetings, for example to the CIPM, in 2006, but cannot be found in a peer-reviewed publication until 2009 (Taylor, 2009). That publication (along with Mohr, 2008) also referred to the proposals as being for a "quantum-based SI", which was explained as being "based on natural physical phenomena, thereby providing a stable and universally reproducible system" (Mohr, 2008).

The 2006 proposals were very detailed. In addition to advocating the principles motivating the changes, the publication proposed how new definitions for the base units might be worded. This is not a simple matter, since a definition based on a fixed value of a constant is most easily articulated using wording that explicitly states the value of the constant from which the unit may be realized. This has subsequently become known as an "explicit constant" formulation for a definition. For example, an explicit-constant definition for the kilogram might be:

> The kilogram, unit of mass, is such that the Planck constant is equal to exactly 6.626 068 96 $\times$ 10^{-34} joule second.

By contrast, the same definition can be expressed using an "explicit unit" formulation as:

> The kilogram, unit of mass, is the mass of a body whose de Broglie-Compton frequency is $(299\ 792\ 458)^2/(6.626\ 068\ 96 \times 10^{-34})$ hertz.

This formulation evokes a frequency of the order 10^{50} Hz, which is vastly larger than any frequency that is encountered routinely and is therefore not preferred.

This choice of a formulation for the definitions that emphasizes the fixed constant rather than the fixed unit has had an influence on the way the proposals have been presented. The increased emphasis on constants leads to the system being presented in such a way that the set of constants plays a more central role than the base units themselves. Indeed, this set of constants is now referred to as the "defining constants" of the SI.

At this point it is useful to consider the nature of these seven defining constants. They are:

i   the frequency of the hyperfine splitting in caesium, which is a property of an atom, that has been the basis of the definition of the second since 1967;

ii   the luminous efficacy of monochromatic radiation at a specified wavelength, which is the value chosen to be the basis of the definition of the candela in 1979;

iii   the speed of light, a constant of nature, that has been the basis of the definition of the metre since 1983;

iv & v   the Planck constant (h) and the electric charge (e), which are fixed in order to fix the numerical values of the Josephson constant (K_J) and the von Klitzing constant (R_K), with the consequence of redefining the kilogram with reference to the fixed numerical value of h;

vi   the Avogadro constant (N_A), which is fixed in order to redefine the mole as a fixed number of entities;

vii   the Boltzmann constant, which is fixed in order to allow the kelvin to be realized by primary thermometers at any temperature.

We observe that the set of seven defining constants do not all arise from quantum physics. The opportunities to realize the proposed definitions through the use of "quantum physics" are no different to those accessible today – except that the use of the Josephson and the quantum Hall effects will be fully assumed back within the SI. Indeed, most primary thermometers do not depend on "quantum" effects, and the possibility of realizing the mole through "counting" atoms is not truly a "quantum" effect.

The 2006 publication concluded that the proposals could be "formally adopted by the 24th CGPM in 2011". However, the CGPM did not adopt the new definitions at its meeting in 2011 but did go so far as to recognize the possibility of adopting the changes at a future meeting. It adopted a resolution with similar wording to that used in 2007:

> Considering . . . that the NMIs and the BIPM have rightfully expended significant effort . . . by extending the frontiers of metrology so that the SI base units can be defined in terms of the invariants of nature – the fundamental physical constants.
>
> (23rd CGPM 2007 (Res 12); 24th CGPM 2011 (Res 1))

They indicated an intention, but they did not explain the motivation for such a change in detail.

5   Articulating the motivation for change

The technical advantages and disadvantages of the proposed changes to the SI described in the previous section are subtle. They cannot be explained simply and would only concern those laboratories working with the SI units at the highest levels of accuracy. It is difficult to generate widespread support to motivate the adoption of complex technical ideas like these. Indeed, the process of balancing technical advantages against technical disadvantages often sets the interests of one group of users against those of another. In some cases, one group of users might be disadvantaged by an aspect of the proposals that provides a benefit to another. For example, in the case considered here, those involved in mass measurements may be disadvantaged at present by the lack of an independent means to validate the performance of the unique IPK artefact. In this respect the redefinition brings an advantage to them. However, the proposed change would allocate an additional fixed uncertainty to all measurement results made traceable to the IPK. Hence a balanced presentation cannot simply describe the proposals as being advantageous to the mass community. (Establishing a utilitarian perspective by which the advantages to one group are offset against the disadvantages to another is beyond the capacity for decision making within global metrology at present.)

   In order to generate widespread support, and sufficient motivation to make such changes, the arguments must be articulated at a level of generality that can be understood widely and are attractive to those bodies that have the authority to endorse them. In this section, we consider some of the more general arguments that have been developed to motivate and justify the adoption of the proposed changes.

5.1   *From artefacts to fundamental physical constants*

The most frequently used of these more general arguments to motivate the change is articulated in terms of the need to eliminate any dependence of the definitions of the base units on physical artefacts. This is often phrased very broadly in terms of the objective "to define the units in such a way that they would be based on natural physical phenomena, thereby providing a stable and universally reproducible system" (Mohr, 2008).

We treat this as a more general argument because, as we illustrate later, it is often presented in a way that goes beyond specific technical issues about the performance of the IPK itself and is articulated at the level of a principle justified axiomatically. For example:

> Although the international prototype has served well as the unit of mass since it was so designated by the CGPM in 1889, it has one important limitation: it is not linked to an invariant of nature. Thus, the possibility of redefining the kilogram in terms of a true natural invariant – the mass of an atom or a fundamental physical constant – has been discussed during at least the last quarter century.
>
> (Mills et al., 2006)

In this statement, we read that the limitation of the IPK is that it is not linked to a natural physical phenomenon or an invariant of nature, not that it is simply unable to fulfil its intended purpose because its performance is inadequate in some practical respect. The argument has become one of principle; it not only addresses the performance of the IPK, but it also addresses its mode of operation.

The use of artefacts to act as standards is a recurring theme in the history of measurement. The signature of the Metre Convention by 17 nations in 1875 established an international agreement that unique artefacts should be carefully prepared to form the basis for the definitions of the metre and the kilogram. These artefacts were formally endorsed by the CGPM in 1889. Whilst the use of these artefacts to define measurement units provided the practical basis for the world's measurement system, even at that time it had been observed that a more stable and enduring basis might be found. This is referred to directly in the widely-referenced quotations:

> If then, we wish to obtain standards of length, time and mass which shall be absolutely permanent, we must seek them not in the dimensions, or the motion, or the mass of our planet, but in the wavelength, the period of vibration, and the absolute mass of these imperishable and unalterable and perfectly similar molecules.
>
> (Maxwell, 1870, p. 225)

and

> With the help of fundamental constants we have the possibility of establishing units of length, time, mass and temperature, which necessarily retain their validity for all times and cultures, even extraterrestrial and nonhuman.
>
> (Planck, 1900)

Great progress was made during the twentieth century towards exploiting the stable and enduring properties of atoms to act as measurement standards. The most prominent examples were:

i the metre, which was defined in 1960 as a specified number of wavelengths in vacuum of a certain transition of the krypton-86 atom; and

ii the second, which was defined in 1967 as the duration of a specified number of periods of the transition between two hyperfine levels in the ground state of the caesium-133 atom.

Subsequently, in 1983 a definition was agreed for the metre that introduced a fundamental constant into the SI when the metre was defined in terms of the speed of light. The use of the speed of light for the definition of the metre is highly effective. There is a link between distance and velocity that makes it intuitive and straightforward to explain. Also, the speed of light plays a fundamental role in the theories of electromagnetism and special relativity. Hence the 1984 definition of the metre is both straightforward to articulate and is based on an important principle from modern physics.

The argument for the elimination of artefacts from the definitions of the base units now focuses on the one remaining unique artefact, the IPK which has been used since 1889 as the definition of the kilogram. Drawbacks that can be rationally associated with it include its vulnerability to accidental damage and the possibility that the alloy itself might be intrinsically unstable, perhaps because of effects on its surface. Such arguments should not be accepted without scrutiny:

> There is clear experimental evidence that the masses of 1 kg prototypes are not stable when compared among themselves over many dozens of years. To date, however, there is no experimental evidence to support a hypothesis that the mass of the international prototype is changing with respect to a mass derived from a fundamental constant of physics. The current lack of such evidence is presumably due to the present level of precision of watt balance and Avogadro experiments.
>
> (Davis, 2005)

Some evidence has become available recently. The results of the extraordinary campaign of calibration of the prototype mass standards held at the BIPM in 2014 show that the difference between the mass of the *témoins* and the IPK has only changed by an average of 1 μg since 1989 (Stock, Barat, Davis, Picard, & Milton, 2015). This sets a significantly different quantitative limit on the practical stability of the IPK as an artefact than the upper limit proposed for its drift, which has been discussed earlier.

However, the argument has also been worded more strongly and at a general level:

> In the 21st century, why should a piece of Pt – Ir alloy forged in the 19th century that sits in a vault in Sevres restrict our knowledge of the values of h and m_e?
>
> (Mills et al., 2006)

Some authors have extended the argument for the elimination of artefacts from the definitions of the base units to include artefacts that are not unique, such as water, which is an essential part of the definition of the kelvin. This is presented as being unsatisfactory because it is difficult to prepare reliably with a suitable isotopic composition to realize the triple point precisely. The argument can be taken to an extreme by extension to caesium (in the definition of the second) and carbon 12 (in the definition of the mole (Davis & Milton, 2014) being considered unsatisfactory simply because they are physical materials that are not intrinsically fundamental. This can bring the argument to the point of questioning whether the property of an atom is sufficiently fundamental.

The alternative to the use of artefacts is to base measurement standards on "constants of nature" or "fundamental constants". This view is usually referenced to the previous quotation from Maxwell, although at least one contemporary author proposes that fundamental constants should be preferred to atomic properties:

> The question of whether the mass of an atom or an elementary particle is more fundamental than the Planck constant has been raised. Since there is a spectrum of particles, each one having a different mass and none more fundamental than the other, giving a fixed value to the Planck constant is surely less arbitrary than giving a fixed value to the mass of an atom.
>
> (Becker et al., 2007)

5.2   *Improving the accuracy of the fundamental constants*

Before progressing to a discussion of the second general argument, for completeness, we consider an argument that was articulated in 2005 but is now used less often:

> Redefining the SI base units kilogram, ampere, kelvin and mole . . . as we have proposed in this paper would implement CIPM Recommendation 1 (CI-2005) in a way that would be profoundly beneficial to both metrology and our knowledge of the values of the fundamental physical constants in SI units, or more generally, to quantum physics.
>
> (Mills et al., 2006)

And from another author:

> There will be a benefit to physicists engaged in high precision experiments as the value of the constants fixed within the SI and combinations of their values will have zero uncertainty and will not change.
>
> (Robinson, 2013)

The counter to these arguments is that it has to be recognized that additional quantities (or constants) with uncertainty must be introduced into the system of fundamental constants when adopting definitions based on fixed values of certain other constants. Hence such claims are not absolute (Cabiati & Bich, 2009; Milton, Williams, & Forbes, 2010) in the sense that whilst some constants will benefit from having a decreased (or even zero) uncertainty, others that were formerly fixed will have an uncertainty introduced where there was none before.

5.3   *Widening access to the realization of the base units*

There is a second group of arguments in favour of the 2006 proposals that was also referred to in Mills et al. (2006). These centre on the opportunity to provide greater access to primary realizations of the base units. For example:

> The most significant benefit in our view is that it liberates mass metrology from an artefact-based unit. This means that different laboratories can realize the unit at will – the long-sought goal of the kilogram being realizable at any time at any place by anyone with the requisite uncertainty will now be limited only by the financial and/or human resources available at a given laboratory.
> (Mills et al., 2006)

This view has been repeated more recently:

> We have but one realization of the unit of mass in the entire world (one mass to rule them all!). If redefinition is successful, we will immediately have at least three {laboratory} groups capable of realizing the unit and calibrating artifacts for use as primary standards. . . . It will be possible to derive a unit of mass using whatever means is convenient and at whatever scale point desired.
> (Pratt, 2014)

And also, with more caution:

> Whilst "quantum" phenomena are, in principle, accessible "for all time to all people", in practice the experimentation required to realize them is not. In the most extreme cases, they depend on experimentation that can only be carried out by highly specialized laboratories with the necessary experience and funding to do so.
> (Milton et al., 2007)

It is important to recognize that such universality of access can also be achieved through classical experiments (for example, the present definition of the kelvin provides universality of access) as much as it can be by a quantum definition. Indeed, a new definition may not be required, for example a watt balance could be used under the present definition of the ampere to realize a measurement of the kilogram. It would give a result in terms of the 1990 convention for the electrical units that are readily convertible to the SI. The extent of access to the realization would be the same.

When the introduction of a metric system was advocated at the time of the French Revolution, we note that its motivation was, at least in part, based on the notion of facilitating freedom and equality of access to measurement standards. However, such ambitions have proved difficult to achieve in practice. Attempts to adopt a definition based on the size of the Earth that had been proposed in order to allow universal access did not achieve their objective. The following has been observed:

> The great virtue of the meridian experiment . . . was that it could not easily be repeated – as a simple pendulum experiment might have been. The meridian experiment, by its very grandeur, difficulty and expense, had fixed the meter – permanently.
>
> (Alder, 2002)

The challenge of achieving the axiom of universality of access can be limited by the practical.

Returning to the motivation for metrology outlined in the introduction to this chapter, it is a pragmatic enterprise that succeeds when it is able to provide measurement results that are stable, comparable and coherent. It is improvements in these three respects that should motivate improvements to the system. Appeals to the need for a system based on stronger fundamental principles that are supported by unstated axioms do not stand unless there continue to be pragmatic benefits. By contrast, increased access to the system is an objective that is motivated pragmatically and is consistent with an accepted axiom. As one of the earliest advocates of quantum metrology wrote:

> The most important property of a system of measurement is that it should be widely and easily available.
>
> (Cook, 1972)

6 Conclusions

Recently, substantial progress has been made towards demonstrating the equivalence of measurements of h by the watt balance and the Si-XRCD methods (Karshenboim, Mohr, & Newell, 2015; Mohr, Newell, Taylor, & Tiesinga, 2018). It is therefore very likely that the proposals discussed here will be adopted at the 26th meeting of the CGPM in 2018. As we have discussed, the most significant consequences of the proposed changes will be:

i   that practical electrical measurements will be brought into the SI (but at the cost of step changes when practical electrical measurements move from the conventional values to the SI values of R_K and K_J at the time of implementation);

ii the opening of the possibility of using any method that can relate a mass measurement to the defined value of h to realize the kilogram (for example the watt balance or the XRCD method); and

iii   the opening of the possibility of using primary thermometric methods to make measurements of temperature in the SI unit kelvin.

In each of these phrases we have chosen to articulate the proposals in terms of the impact they will have on users. However, as discussed in this chapter, articulating the motivation for the changes in this way may not have been sufficiently compelling to gain widespread support, particularly in view of the additional complexity incurred in formulating the definition of the kilogram. In contrast, emphasizing the benefits of adopting definitions that refer to "invariants of nature" has drawn attention to the firm fundamental basis for the definitions, and it may have deflected attention from the practical challenges of realizing them. Examples include the following statements:

> The value of the SI as a common measurement language for intelligible communication between practical metrology and quantum physics will be significantly improved.
>
> (Mills et al., 2006)

> Above all, the changes in the SI proposed here will strengthen the philosophical foundation of our system of units in relation to our present understanding of theoretical and quantum physics.
>
> (Mills, Mohr, Quinn, Taylor, & Williams, 2011)

In conclusion, we note that following the proposed changes, any laboratory will, in principle, be able to realize the definitions and the scope of methods allowed for their realization will increase. In this respect, the proposed changes may bring the SI closer to one of the greatest ambitions of measurement science – that it should provide access to the basis of measurements "for all people for all time", a statement that goes back to the law of 1799 that initiated the first metric system in France.

As we have discussed, the 2006 proposals will not implement a complete "quantum SI" because the system will continue to be dependent on practical (and classical) methods for its implementation, at least until future developments in quantum metrology of comparable significance to the discovery of the Josephson and von Klitzing effects. Re-organizing and re-wording definitions, as has been done for the metre, second and candela, does not of itself make them more dependent on "invariants of nature" than they were before.

We should not expect the proposed changes to be the end of the process of improving the international system of measurement units. It is to be expected that further progress will be made in metrology, eventually leading to proposals to revise the "New SI", not least to incorporate a definition of the second based on an optical transition in place of the present microwave transition. This would be a change within the framework of a system, in which the realization of all of the base units is already open "for all people for all time". This aspect of the present change will be one of historical importance. There should be no reticence about this. It was highlighted by the authors of the 2006 proposals as being "the most significant benefit". It is a point that may have been inadvertently diminished by the emphasis on "defining constants" and "invariants of nature", to the point where the CGPM has not touched upon it in any of its three recommendations on the subject.

References

Alder, K. (2002). *The measure of all things*. London: Abacus.

Becker, P., de Bièvre, P., Fujii, K., Gläser, M., Inglis, B., Luebbig, H., & Mana, G. (2007). Considerations on future redefinitions of the kilogram, the mole and of other units. *Metrologia, 44*, 1–14.

BIPM (2016). *Bureau International des Poids et Mesures Strategic Plan* (2018). France: BIPM.

Cabiati, F., & Bich, W. (2009). Thoughts on a changing SI. *Metrologia, 46*, 457–466.

Cook, A. H. (1972). Quantum metrology: Standards of measurement based on atomic and quantum phenomena. *Reports on Progress in Physics, 35*(2), 463–528.

Davis, R. S. (2005). Possible new definitions of the kilogram. *Philosophical Transactions of the Royal Society A: Mathematical, Physical and Engineering Sciences, 363*, 2249–2264.

Davis, R., & Milton, M. J. T. (2014). The assumption of the conservation of mass and its implications for present and future definitions of the kilogram and the mole. *Metrologia, 51*, 169–173.

Fletcher, N., Rietveld, G., Olthoff, J., Budovsky, I., & Milton, M. (2014). Electrical units in the new SI: Saying goodbye to the 1990 values. *Measurement, 9*, 30–35.

Gläser, M., Borys, M., Ratschko, D., & Schwartz, R. (2010). Redefinition of the kilogram and the impact on its future dissemination. *Metrologia, 47*, 419–428. https://doi.org/10.1088/0026-1394/47/4/007.

Karshenboim, S. G., Mohr, P. J., & Newell, D. B. (2015). Advances in determination of fundamental constants. *Journal of Physical and Chemical Reference Data, 44*, 031101.

Maxwell, J. C. (1870). *Address to the Mathematical and Physical Sections of the British Association, Liverpool, 15 September*. British Association Report, 40, reproduced in The Scientific Papers of James Clerk Maxwell (Vol. 2), ed. W. D. Niven. Cambridge: Cambridge University Press, 1890.

Mills, I. M., Mohr, P. J., Quinn, T. J., Taylor, B. N., & Williams, E. R. (2005). Redefinition of the kilogram: A decision whose time has come. *Metrologia, 42*, 71–80.

Mills, I. M., Mohr, P. J., Quinn, T. J., Taylor, B. N., & Williams, E. R. (2006). Redefinition of the kilogram, ampere, kelvin and mole: A proposed approach to implementing CIPM recommendation 1 (CI-2005). *Metrologia, 43*, 227–246.

Mills, I. M., Mohr, P. J., Quinn, T. J., Taylor, B. N., & Williams, E. R. (2011). Adapting the International System of Units to the twenty-first century. *Philosophical Transactions of the Royal Society, 369*, 3907–3924.

Milton, M. J. T. (2013). The mole, amount of substance and primary methods. *Metrologia, 50*, 158–163.

Milton, M. J. T., Davis, R., & Fletcher, N. (2014). Towards a new SI: A review of progress made since 2011. *Metrologia, 51*, R21–R30.

Milton, M. J. T., & Mills, I. M. (2009). Amount of substance and the proposed redefinition of the mole. *Metrologia, 46,* 332–338.

Milton, M. J. T., Williams, J. M., & Bennett, S. J. (2007). Modernizing the SI: Towards an improved, accessible and enduring system. *Metrologia, 44,* 356–364.

Milton, M. J. T., Williams, J. M., & Forbes, A. B. (2010). The quantum metrology triangle and the redefinition of the SI ampere and kilogram; analysis of a reduced set of observational equations. *Metrologia, 47,* 279–286.

Mohr, P. J. (2008). Defining units in the quantum based SI. *Metrologia, 45,* 129–133.

Mohr, P. J., Newell, D. B., Taylor, B. N., & Tiesinga, E. (2018). Data and analysis for the CODATA 2017 special fundamental constants adjustment. *Metrologia, 55,* 125–146s. https://doi.org/10.1088/1681-7575/aa99bc

NBS. (1971, July). *A metric America: A decision whose time has come.* Special Publication 345, Washington, DC: National Bureau of Standards.

Planck, M. (1900). cited in R. P. Huebner and H. Luebbig. *A Focus for discoveries* (Singapore: World Scientific, 2008), p. 150.

Pratt, J. R. (2014). How to weigh everything from atoms to apples using the revised SI. *Measurement, 9,* 26.

Robinson, I. A. (2013). Planck, Avogadro and measuring mass using fundamental constants. *Annals of Physics* (Berlin), *525*(8–9), A135–A137.

Stock, M., Barat, P., Davis, R. S., Picard, A., & Milton, M. J. T. (2015). Calibration campaign against the international prototype of the kilogram in anticipation of the redefinition of the kilogram part I: Comparison of the international prototype with its official copies. *Metrologia, 52,* 310–316.

Taylor, B. N. (2009). Molar mass and related quantities in the New SI. *Metrologia, 46,* L16–L19.

Taylor, B. N., & Mohr, P. J. (1999). On the redefinition of the kilogram. *Metrologia, 36,* 63–64.

3 Reforming the International System of Units

On our way to redefine the base units solely from fundamental constants and beyond

Christian Bordé

1 Introduction: from the French Revolution to Max Planck

The metric system was born during the French Revolution with the idea of settling a universal system of units, open to every people, in every time. At that time, the dimensions of the Earth and the properties of water seemed to offer a universal basis, but sometime later, James Clerk Maxwell judged them less universal than the properties of the molecules themselves. The next step was taken by George Johnstone-Stoney, then by Max Planck, showing that a deeper aim was to found the system of units only on a set of fundamental constants originating from theoretical physics. Consequently, a long divorce began between the practical requirements of instrumental metrology and the dreams of theoretical physicists. Might they marry again? This has now become possible thanks to a set of recent discoveries and new technologies: laser measurements of length, Josephson effect, quantum Hall effect, cold atoms, atom interferometry, optical clocks, optical frequency measurements. . . . So a strong tendency to tie the base units to fundamental constants is rising again, and the debate is open as to the relevance, the opportunity and the formulation of new definitions.

Since the very beginning of this adventure the French Academy of Sciences has had a leading role in the development of ideas and the settling of the metric system. An Academy Committee on Science and Metrology is still working on this theme today. I outline in this chapter some of the questions under consideration in this committee and their underlying physical grounds. My purpose is to offer a logical analysis of the system of units and to explore possible paths towards a consistent and unified system with an original perspective. The path taken here builds on the fact that, thanks to modern quantum technologies, any measurement can be reduced to a dimensionless phase measurement due to optical or matter-wave interferometry, and I shall try to follow this simple guiding line. I shall finally show how one could progress even further on the path of a synthetic framework for fundamental metrology based upon pure geometry in five dimensions. The reader who does not wish to enter into these mathematical considerations can skip the last paragraph and Appendix 2. The conclusion emphasizes the role of quantum mechanics and uncertainty relations in the new metrology.

2   Present status and evolution of the International System of Units (the SI)

The SI (11th CGPM, 1960) comprises seven base units which are all more or less concerned with the process of evolution mentioned earlier:

i   the metre has already been given a new definition from the time unit and the velocity of light in 1983 (see next section);

ii   the kilogram is still defined today by an artefact of iridium/platinum alloy, but as we shall see in detail it could find a new definition from Planck's constant in a near future;

iii   the SI ampere is defined through a property of the vacuum, specifically its magnetic permeability $\mu_0 = 4\pi.10^{-7}$ H/m (9th CGPM, 1948), but the electrical units have *de facto* already gained their independence from the SI ampere by adopting conventional values for Josephson and von Klitzing constants, and the natural temptation today is to adopt the value of the electric charge e in order to freeze the numerical values of these constants;

iv   the kelvin is defined through the triple point of water, whereas fixing Boltzmann's constant k_B would be more satisfactory;

v   the candela, unit of luminous intensity, is nothing else but a physiological unit derived from an energy flux, hence it is redundant with the other base units. Furthermore, it does not take into account the coherence properties, spectral content and spatial mode content of the source. Hence, we shall not give any further consideration to this pseudo-base unit;

vi   the mole (added to the SI by the 14th CGPM in 1971) is defined from the mass of the carbon atom by a dimensionless number, the Avogadro constant. A better determination of this number should give an alternative option in which it would be fixed to redefine the mass unit from the mass of an atom or of the electron. The tendency today is simply to retain this numerical value as a conventional number of entities;

vii   the second, a unit of time, was originally defined as the fraction 1/86 400 of the "mean solar day". The exact definition of the "mean solar day" was left to astronomers. However, observations have shown that this definition was not satisfactory owing to irregularities in the Earth's rotation. To give more accuracy to the definition, the 11th CGPM (1960) approved a definition given by the International Astronomical Union based on the tropical year 1900. But experimental work had already shown that an atomic standard of time, based on a transition between two energy levels of an atom or a molecule, could be realized and reproduced much more accurately. Considering that a more precise definition of the unit of time was essential for science and technology, the 13th CGPM (1967/68) replaced the definition of the second by the following (source BIPM):

> The second is the duration of 9 192 631 770 periods of the radiation corresponding to the transition between the two hyperfine levels of the ground state of the caesium 133 atom. It follows that the hyperfine splitting in the ground state of the caesium 133 atom is exactly 9 192 631 770 hertz.

This definition refers to a caesium atom at rest at a temperature of 0 K. This implies that the corresponding frequency should be corrected from Doppler shifts and from the shifts coming from all sources of ambient radiation (CCTF 1999).

The second should soon be better defined by an optical clock or even by a much higher frequency clock (nuclear transition or matter-antimatter annihilation process). Ideally, physicists would have dreamed of an atomic hydrogen clock; it would have allowed to tie the time unit to the Rydberg constant and possibly, someday, to the electron mass. But this choice could be behind us today.

One should emphasize that the unit of time refers to proper time.[1] Proper time, as we shall discuss in detail, is associated with the internal evolution of a massive object such as an atom and is the measurable quantity from which the time coordinate is more or less artificially constructed. The time unit definition should thus have referred to the atom Bohr frequency and not to the radiation frequency. Among other things no mention is made of the recoil shift.

So, we are facing ill-assorted definitions piled up along the years, without any global consistency. The direct connection between the definition of a base unit from a fundamental constant, the practical working out of it and a main scientific discovery is well illustrated by the case of the metre and its new definition issued from the technological progress of laser sources. This represents the archetype of the path to be followed for the other units.

3   The example of the metre

The metre is the best-known example of a base unit for which a new definition was based upon a fundamental constant, the velocity of light in vacuum c, thanks to progress in optics and especially laser physics during the second half of the twentieth century.

The coordinates of space and time are naturally connected by Lorentz transformations within the conceptual frame of the theory of relativity, and the velocity of light takes place as a factor of conversion in these symmetry transformations. The existence of a symmetry is a first situation that allows us to create an association between two units, and hence to reduce the number of independent units.

A second favourable condition is the existence of mature technologies to implement this symmetry. Relativity uses clocks and rods to define time and space coordinates. The rods of relativity are totally based on the propagation properties of light waves, either in the form of light pulses or of continuous beams whose frequency can now be locked to atomic clocks. It was possible to redefine the length unit from the time unit because modern optics allowed not only the measurement of the speed of light generated by superstable lasers with a relative uncertainty lower than the best length measurements, but also because today the same techniques allow the new definition of the metre to be realized in an easy and daily way.

It is precisely optical interferometry and especially the work of Albert A. Michelson (Nobel Prize winner in 1907) that allow us to go from the nanometric length, which is the wavelength linked to an atomic transition, to a macroscopic length at the metre level. Michelson interferometers can measure the tiniest length variations (10^{-23}) induced by gravitational waves over distances ranging between hundreds of kilometres on Earth and millions of kilometres in spatial projects, such as LISA. Any length measurement can thus be reduced to a phase measurement, i.e. to the determination of an invariant number.

This evolution started in 1960 when the metre was redefined from the radiation of the krypton lamp. The birth of lasers, in 1959, helped to carry on steadfastly in that direction. Above all it was the discovery of sub-Doppler spectroscopic methods, and particularly of saturated absorption spectroscopy in 1969 (Barger & Hall, 1969; Bordé, 1970), which turned lasers into sources of stable and reproducible optical frequencies. The other revolution was the technique of the MIM diodes introduced by Ali Javan that led to measuring the frequency of these light sources directly from the caesium clock. From then on, the velocity of light could be measured with a sufficiently small uncertainty, and so the CGPM in 1983 fixed its value linking the metre to the second. These developments imply a procedure to put the definition into practice (*mise en pratique*), using wavelengths of lasers locked to recommended atomic or molecular transitions.

Finally, this redefinition was possible because there was a theoretical background universally accepted to describe the propagation of light in real interferometers.

To extend this approach, let us investigate to what extent a similar situation can be met for the other units and what fundamental constants are available for each of them. A detailed discussion, partly reproduced in Appendix 1, is given in Bordé (2004, 2005).

4   The dimensioned and dimensionless fundamental constants and their place in present physics

The fundamental constants we are referring to come out of the major theories of modern physics: relativity theory, quantum mechanics, statistical mechanics, field theories, etc. Consequently, they rely on our models and representations of the physical world.

What set of fundamental constants must we choose in the end? They belong to two very distinct categories. On the one hand, we have what can be called conversion constants. Such constants are used to connect quantities originally believed to be of a different nature, but later understood to refer to the same physical entity. A famous example is the equivalence between heat and work that led to the mechanical equivalent of the calorie: 4.18 joules. The conversion constants have the dimension of the ratio between the linked units. They can be given a fixed numerical value, and the number of independent units is thus reduced. Several constants play this role unequivocally: such was the case with the velocity of light, and it is still the case with Planck and Boltzmann constants as discussed later. In other cases, we will have a choice to make between several constants of the same nature: it will be the case of the electric charge, for instance. On the other hand, nature forces on us another sort of constants: the value of non-dimensional ratios. Such are, for example, the coupling constants linked to the fundamental interactions. The best known are the fine structure constant describing the coupling of matter with the electromagnetic field:

$$\alpha = \frac{\mu_0 c e^2}{4\pi\hbar} \tag{1}$$

and its gravitational analogue

$$\alpha_G = G m_e^2 / \hbar c \tag{2}$$

involving the gravitation constant G and the electron mass m_e.

The value of these coupling constants cannot be discussed and remains independent of the system of units. It is a constraint to be considered in our choices.

5   The kilogram and the mole: determining the Avogadro constant with the silicon sphere

Since 1889 (1rst CGPM), the mass unit has been the mass of the international prototype, a platinum-iridium alloy cylinder baptized $\mathfrak{K}$ and kept in a vault of the Pavillon de Breteuil with six copies. After the three intercomparisons made in 1889, 1946/53 and 1989/92, there is now a general agreement on the idea that the mass of the standard prototype, constant by definition, has in fact drifted by several 10 or so micrograms (i.e. some 10^{-8} in relative value). This situation, in which the electrons and other elementary particles of the universe have a mass value changing with time, when the piece of metal in the vault in Sèvres has not, is quite embarrassing. So, every effort must be done to modify the definition (recommendation of the 21rst CGPM). It would be much more satisfactory and justified to start from the mass of microscopic particles (electron or atom) *a priori* quite reproducible, and then to climb up the macroscopic scale. But if masses can be easily compared both at the macroscopic and at the atomic scales, the connection between these two scales is quite difficult. To make this connection we need to make an object with a known number of atoms and whose mass could be compared to that of the standard kilogram. This amounts to determining the Avogadro constant N_A which defines the mole. The mole is a quantity of microscopic objects defined as a conventional number of identical objects. This number (of course without dimension) has been arbitrarily chosen equal to the number of supposedly isolated atoms, at rest and in their fundamental state, contained in 0.012 kg of carbon 12. Consequently, it is, up to a numerical factor 0.012, the ratio of the mass of the standard prototype to the mass of a carbon atom. Avogadro's constant

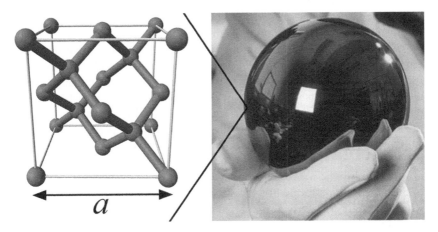

Figure 3.1 The silicon sphere method. Starting with a silicon monocrystal purified by the floating zone method, several nearly perfect spheres with masses ~ 1 kg were made (surface defects below some tens of nanometres) then, thanks to mass spectrometry, X and optical interferometries, the size a of the cell d_{220}, the density $\rho = V/m$ and the molar mass M were determined. The cubic crystal cell corresponding to eight atoms it was possible to obtain the Avogadro constant from the formula $N_A = 8M/(\rho a^3)$.

N_A generally refers to that same number per mole, and it is expressed in mol^{-1}. This number and this constant are just another way of expressing the mass of a carbon atom, or its twelfth part, which is the unified atomic mass unit m_u.

An international programme (the XRCD programme for X-ray crystal density programme) has been developed to determine the Avogadro constant from the knowledge of a silicon sphere studied under "every angle": physical characteristics of dimension, mass, volume, cell parameter, isotopic composition, surface state, etc. (see Figure 3.1). The International Avogadro Coordination project is refining the application of the XRCD method to isotopically enriched ^{28}Si spheres, with the goal of reaching a 1.5×10^{-8} relative standard uncertainty (Andreas et al., 2011). The Avogadro constant N_A based on these measurements is presently $6.022\ 140\ 76(19) \times 10^{23}\ mol^{-1}$ (Vocke, Rabb, & Turk, 2014). This programme has already faced and overcome many difficulties, and one day it should eventually reach the goal of fixing the Avogadro constant with an accuracy that allows a redefinition of the kilogram.

6 Mass concept from proper time: relativity and quantum mechanics

In fact, the notion of mass does not boil down to that of a quantity of matter, and, if a redefinition of the mass unit from the mass of a reference elementary particle goes the right way, it does not reduce the number of independent units. However, there is a possibility, as in the case of the metre, to link the mass unit to the time unit. Indeed, the theory of relativity allows us to identify the mass m of an object with its internal energy, according to the well-known relation $E = mc^2$. What is more, Louis de Broglie, in his famous Note in 1923, teaches us that this energy can be linked to the proper time τ of the object to produce the phase of an internal oscillation. The product $mc^2\tau$ of these two quantities is an action, which must be related to an elementary action, Planck's constant h, to give the phase without dimension of that oscillation $mc^2\tau/h$ (see Appendix 2). In other words, the quantity mc^2/h is a frequency that we shall call the de Broglie-Compton frequency (dBC).[2] This frequency can be indirectly measured in the case of microscopic particles such as atoms or molecules by modern techniques of atomic interferometry, in which de Broglie waves are precisely made to interfere. The first experiments of this type were performed by measuring the recoil frequency shift that occurs when laser light is absorbed or emitted by molecules in saturation spectroscopy (Hall & Bordé, 1974; Hall, Bordé & Uehara, 1976). They were followed by cold atom interferometry (Young, Kasevich, & Chu, 1997) using Bordé-Ramsey interferometers (Bordé, 1989, 1997; Berman, 1997). Today this measurement is done with a relative uncertainty less than 10^{-8} (Biraben et al. (Cadoret et al., 2008; Bouchendira et al., 2011)). From that point, by simply multiplying with the Avogadro constant $\mathfrak{N}_A$, we can have access to the de Broglie-Compton frequency of the kilogram from that of the atomic mass unit m_u,

$$\nu_{dBC} = \frac{M_{\mathring{\mathrm{K}}}c^2}{h} = 1000\mathfrak{N}_A\left(\frac{m_u c^2}{h}\right) \tag{3}$$

and so link the mass unit to the time unit. Then the mass unit would be defined by fixing that de Broglie-Compton frequency, which amounts to fixing Planck's constant (Wignall, 1992). Such was the recommendation made by the working group of the Académie des Sciences to the CIPM in 2005. The definition of the unit of mass would essentially look like: "The kilogram is the unit of mass, it is the mass of a body whose de Broglie-Compton frequency is equal to $(299\ 792458)^2/(6.6260693.10^{-34})$ hertz exactly". This definition has the effect of fixing the value of the Planck constant, h, to be $6.6260693.10^{-34}$ joule second exactly.

This definition currently meets several criticisms: besides being an unfamiliar concept involving too large a number, it is a quantum-mechanical concept used in a range where its validity may be questioned because of decoherence among other things.[3] There is certainly no physical clock at such a high frequency, which thus appears as fictitious. Even at the single atom level the connection between the de Broglie-Compton frequency, which is measured indirectly, and a real clock frequency has been the subject of recent controversy. We shall see in the generalized 5D approach that the overall action and hence the overall phase cancels along the classical trajectory. Interference fringes result only from the phase added by a coupling of modes with different wave vectors or frequencies. A real atomic clock is generated at the Bohr frequency by a superposition of two internal states b and a, and it oscillates at the difference of the two corresponding de Broglie-Compton frequencies on both sides of an interferometer:

$$\nu_{\text{Bohr}} = \frac{m_b c^2}{h} - \frac{m_a c^2}{h} \tag{4}$$

The unit of mass may now be defined from this difference of the de Broglie-Compton frequencies of both states that has a clear physical signification and, if the chosen transition is the atomic transition which defines the unit of time, we make an explicit link between both units: "The kilogram is the unit of mass, it is the mass of N massive particles without mutual interactions with a mass equal to the mass difference between the two internal states which define the unit of time", where N is a fixed numerical value of $c^2/h\nu_{\text{Bohr}}$ obtained by fixing the value of the Planck constant, h, to be $6.6260693.10^{-34}$ joule second exactly.

We shall come back on this point since, as we will see, another way of measuring the de Broglie-Compton frequency of the kilogram exists; it uses the spectacular progress of quantum electric metrology that we are now going to recall.

7 Quantum electric metrology: Josephson and quantum Hall effects

The electrical units underwent two quantum revolutions at the end of the previous century: the Josephson effect allows us to realize the volt, and the quantum Hall effect allows us to carry out the ohm.

Historically the ampere was the first example, before the metre, of a unit defined from a fundamental constant, the magnetic permeability μ_0 of the vacuum (9th CGPM 1948). The combination of these two definitions fixes all the propagation properties of electromagnetic waves in a vacuum: velocity c and impedance $Z_0 = \mu_0 c$. Let us remark that by fixing Planck's constant, an electric charge would be also fixed, the Planck charge given by:

$$q_P = \sqrt{2h / Z_0} \tag{5}$$

In practice, the reproducibility of Josephson and quantum Hall effects (respectively 10^{-10} and 10^{-9} in relative value) reaches such a level that today's electrical measurements use these effects without any other connection to the definition of the ampere. Were Planck's constant fixed, the electricians would be greatly tempted to fix the electron charge rather than Planck's charge, having in the back of their mind to fix Josephson and von Klitzing's constants. Unfortunately, the simple theoretical expressions that link these two constants to e and h have not yet been validated with high enough accuracy (only 2.10^{-7} for K_J and 3.10^{-8} for R_K in relative value), even if their universality could be demonstrated at a much higher level. Independently from the strong theoretical arguments that lie under these formulas, it is necessary to make sure that possible corrections are low enough for both effects to achieve a reliable real-ization of $2e/h$ and h/e^2 (see Figures 3.2–3.4).

In the case of the quantum Hall effect, such a verification can be made because the ratio of the vacuum impedance to h/e^2 is just the double of the fine structure constant α. The vacuum impedance can be realized thanks to a calculable capaci-tor (Thomson-Lampard), and the comparison of Z_0/R_K with 2α value obtained by atom interferometry presently sets an uncertainty level around 10^{-8} and will

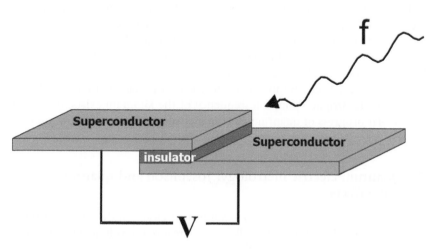

Figure 3.2 Josephson junction.

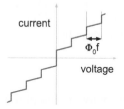

Figure 3.3 The Josephson effect (Nobel Prize 1973) uses the junction comprising a very thin insulating layer sandwiched between two supraconducting plates. When this junction is irradiated by an electromagnetic wave of frequency f, its current-voltage characteristic presents voltage plateaus connected to the frequency f by a simple proportionality relation in which n is an integer characterizing each plateau: $V = nK_J^{-1}f$. The Josephson constant K_J is given with an excellent approximation by $2e/h$. The charge $2e$ is that of Cooper pairs of electrons that are able to tunnel across the junction. This effect has a topological nature ($\phi_0 = h/2e$ is a quantum of flux), hence its universal character, independent of the detailed realization of the junction and veri-fied at the 10^{-10} accuracy level.

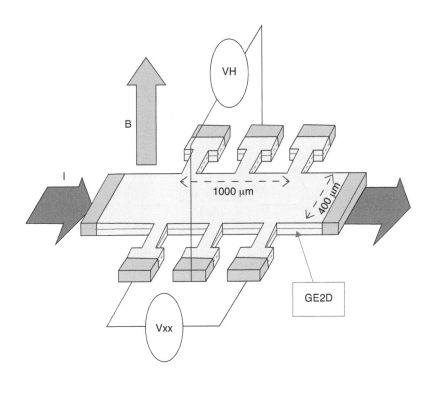

VH

B

I

1000 μm

400 μm

GE2D

Vxx

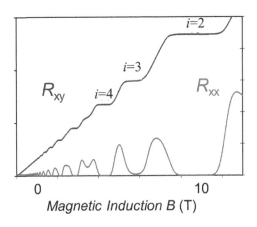

$i=2$

$i=3$

R_{xy}

$i=4$

R_{xx}

0

10

Magnetic Induction B (T)

Figure 3.4 Quantum Hall effect. When a bidimensional gas of electrons in a semicon-
ductor is submitted to a strong magnetic field, the transverse resistance
(Hall resistance) exhibits steps quantized by the integer i and von Klitzing
(Nobel Prize 1985) resistance R_K, whereas the longitudinal resistance vanishes: $R_H = R_K/i$. Here again the effect has a universal topological nature
and is protected by the chiral anomaly introduced by Schwinger, which
suggests that $R_K = h/e^2$ with an excellent approximation.

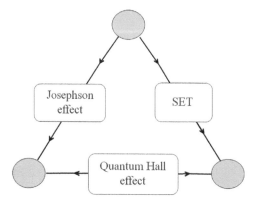

Figure 3.5 The metrological triangle is the quantum realization of Ohm's law. If the three effects are described by the canonical formulas with the same constants e and h in the three formulas: $U = \dfrac{h}{2e} f$, $I = ef'$, $\dfrac{R}{R_K} = \dfrac{e^2}{h} R$ one must check that $U = RI$ leads to an equality between frequencies $f = 2(R/R_K)f'$.

certainly improve beyond 10^{-8}. In the case of the Josephson effect, the limit comes from our insufficient knowledge of the proton gyromagnetic ratio. Fortunately, two other verifications will be possible with the metrologic triangle and the watt balance as we shall see.

In fact, quantum electrical metrology is undergoing a third revolution, with the SET (Single Electron Tunnelling) permitting to count electrons one by one. Then Ohm's law becomes an equality between frequencies: the electric potential difference is turned to a Josephson frequency, the electric current to a number of electrons by second and the electric resistance expressed in terms of von Klitzing resistance has no dimension (see Figure 3.5).

The closure of the metrological triangle (Piquemal et al., 2004) will be a real test of the quantum realizations and of the theories that connect K_J and R_K to the fundamental constants of physics. Presently it is done at some 10^{-7} level, but hopefully that limit will reach the 10^{-8} level in the future.

Thus, electrical metrology is evolving in a profound way. In the future it will occupy a key position for the entire metrology, especially thanks to the "electric" kilogram (see next section). As for the electrical units, the question can be raised if one should fix the positron charge e or rather Planck's charge q_p? The ratio of both charges being the square root of the fine structure constant, the corresponding uncertainty will be transferred to the non-fixed charge. Some arguments inspired by the recent theories of strings and an easier statement of gauge invariance point to the first choice. Caution towards the formulae giving K_J and R_K speaks for the second choice, which goes back to keep the vacuum impedance fixed as it is now. This choice has in fact already been made by the CIPM, and the last recommendations of the CGPM are in favour of fixing the electric charge e.

8   The electric kilogram and the watt balance

If the formulae giving K_J and R_K are valid, Josephson and quantum Hall effects can be combined to produce an electric power proportional to Planck's constant (Figure 3.6):

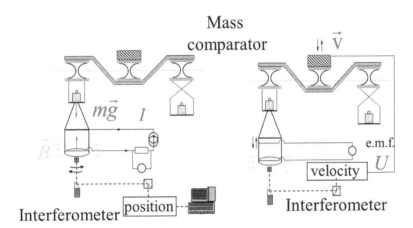

Figure 3.6 The Watt Balance. In its classical version, the watt balance is operated in two steps. In the first one, the weight of the kilogram in the gravity field is balanced by the Laplace force exerted on a coil conducting an electric current and placed in a magnetic field. The current I is measured by the combination of Josephson and quantum Hall effects. In the second step the same coil is moved at constant speed v in the same magnetic field and the induced emf U is measured thanks to the Josephson effect. Under these conditions, the properties of the coil and of the magnetic field are common to both modes, and the final formula expressing the equality between electrical and mechanical powers $mgv = UI$ involves only times and frequencies for the de Broglie-Compton frequency of the kilogram $\nu_{dBC} = M_g c^2/h = f_1 f_2 \omega T^2/[4\varphi(v/c)]$, where the Josephson frequencies of the two steps are f_1 and f_2. The velocity v is measured by optical interferometry and the terrestrial gravity field g by atom interferometry (phase $\varphi = kgT^2$) using laser pulses of pulsation $\omega = kc$ separated by the time interval T.

$$UI = (h/2e)^2 / (h/e^2) f_1 f_2 = \frac{h}{4} f_1 f_2 \qquad (6)$$

where f_1 and f_2 are the two Josephson frequencies involved in this measurement.

   This opened a new way of measuring the de Broglie-Compton frequency of the kilogram, that of the "electric" kilogram. The "electric kilogram" was born with the watt balance, suggested by Kibble in 1975, which in one step (cryogenic version of the BIPM) or two steps (see Figure 3.6) carries out the direct comparison between a mechanical watt realized by moving a mass in the gravitational field of the Earth, and an electric watt given by the combination of Josephson and quantum Hall effects. Such a method demonstrated more than 20 years ago in the USA and in Great Britain that it could reach a level of relative uncertainty consistent with that of the present kilogram, i.e. some 10^{-8}. Two new realizations have been assembled and are under study, one in Switzerland and a newer one in France. Other programmes will follow. Within a few years this effort is likely to offer the opportunity to keep track of the evolution of the present kilogram prototype, and later to give a new definition of this kilogram by fixing its dBC frequency. Clearly there is a competition between two projects to define the mass unit: in the first project Planck's constant is fixed, and the watt balance allows us to measure masses easily; in the second one, Avogadro's constant is determined and fixed, and the mass unit is defined from an elementary mass such as the electron mass. However, in that case, the practical realization of a macroscopic mass still must be done through the realization of a macroscopic object whose number of microscopic entities is known. The first point of view is the most attractive on the conceptual, theoretical and even practical levels, even if the mass unit expression is not easy to grasp for everyone. Anyway, both ways towards Planck's constant will need to be fully reconciled.

The CCM[4] (2013) therefore *recommends* that the following conditions be met before the CIPM[5] asks CODATA[6] to adjust the values of the fundamental physical constants from which a fixed numerical value of the Planck constant will be adopted:

i   at least three independent experiments, including work from watt balance and XRCD experiments, yield consistent values of the Planck constant with relative standard uncertainties not larger than 5 parts in 10^8;

ii   at least one of these results should have a relative standard uncertainty not larger than 2 parts in 10^8; and

iii the BIPM prototypes, the BIPM ensemble of reference mass standards used in the watt balance and XRCD experiments, have been compared as directly as possible with the international prototype of the kilogram.

There has been much progress in the determination of Planck's constant in the recent past. A recent evaluation at NIST resulted in a value published in 2014 with a standard uncertainty of 4.5 parts in 10^8. The last result from NRC has a standard uncertainty of 1.8 parts in 10^8, enough to meet condition 2 of CCM. There are a number of other watt balance experiments that will provide independent values.

We may imagine future ideal versions of the watt balance working as true matter-wave interferometers analogous to superconducting ring gyros (Zimmermann & Mercereau, 1965; Parker & Simmonds, 1971). A superconducting coil directly connected to a Josephson junction works as a Cooper pair interferometer which experiences a phase shift from the change in gravitational potential. The power balance

$$M_{g}gv = IU \qquad\qquad (7)$$

may be written as

$$M_{g}c^{2}\left(\frac{gvT}{c^{2}}\right) = N_{2e}hf_{J} \tag{8}$$

in which $2eN_{2e}/T$ is the intensity of supercurrent, and v and T are respectively the velocity and the duration of the coil vertical motion. The Josephson frequency f_{J} thus appears as a measurement of the gravitational shift of the dBC frequency of N_{2e} Cooper pairs sharing the mass M_{g}.

One can have a fascinating discussion on the question of whether quantum mechanics applies or not when at the macroscopic scale of the kilogram and on the real significance of the appearance of the Planck constant in the watt balance formula. This debate has already started and must continue. Whatever comes out, we are all already persuaded that the mass of a macroscopic object is the sum of that of all its microscopic constituents and of a weak approximately calculable interaction term. This hypothesis is implicit in both possible new definitions of the unit of mass. The concept of mass must be identical at all scales and mass is an additive quantity in the nonrelativistic limit. There is no doubt also that, at the atomic scale, mass is directly associated with a frequency via the Planck constant. This frequency can be measured for atoms and molecules even though it is quite a large frequency. As mentioned earlier, measurements of mc^{2}/h are presently performed with a relative uncertainty much better than 10^{-8}. By additivity, the link between a macroscopic mass and a frequency is thus unavoidable. If one accepts to redefine the unit of mass from that of a microscopic particle, such as the electron, then the link with the unit of time is *ipso facto* established with a relative uncertainty much better than 10^{-8}.

Both units are *de facto* linked by the Planck constant to better than 10^{-8}. It seems difficult to ignore this link and not to inscribe it in the formulation of the system of units, especially since it leads to a reduction of the number of independent units.

Another extremely important point is that mass is a relativistic invariant. It should thus never be associated with the frequency of a photon field, which transforms as the time component of a 4-vector in reference frame changes. The de Broglie-Compton frequency is a proper frequency, Lorentz scalar, equal by definition to mc^{2}/h.

Last, in the hypothesis of the mass unit being redefined by fixing Planck's constant, the mole could be redefined separately from the kilogram by fixing Avogadro's constant. But should we not keep an exact molecular mass for carbon 12? If the mole is not any more directly connected to 12 grams of carbon, its definition amounts to define an arbitrary number, and this number cannot be considered as a fundamental constant of nature. It is only if the mole remains defined by 12 grams of carbon that it rests on a true physical constant, the mass of the carbon atom. This constant has to be determined experimentally if the unit of mass is defined by fixing the Planck constant.

9   Boltzmann's constant and the temperature unit

Statistical mechanics allows us to go from probability to entropy, thanks to another dimensioned fundamental constant, Boltzmann's constant k_B. Presently the scale of temperature is arbitrarily fixed by the water triple point, a natural phenomenon of course, but yet very far from fundamental constants.

By analogy with the case of Planck's constant, it seems natural to propose fixing Boltzmann's constant k_B. Indeed, there is a deep analogy between the two "S's" of physics, which are action and entropy. They provide respectively the phases and the amplitudes for the density operator. The corresponding conjugate variables of energy are time and reciprocal temperature with the two associated fundamental constants: the quantum of action h and the quantum of information k_B. Both participate in statistical quantum mechanics through their ratio k_B/h. The evolution parameter θ that comes in naturally[7] in the combination of Liouville-von Neumann and Bloch equations for the density operator ρ is the complex time:

$$\theta = t + i\hbar\beta/2 = t + i\hbar/2k_B T \tag{9}$$

The link between atom interferometry and the Doppler broadening of line shapes by the thermal motion of atoms is established in reference (Bordé, 2009), which brings the connection between phase and temperature measurements. The thermal motion of atoms is responsible for a loss of phase coherence, and the Doppler broadening may be seen as a limited visibility of interference fringes.

An interesting analogy may be drawn for the two inaccessible limits that are the velocity of light c and the absolute zero temperature $T = 0$. In both cases the corresponding variable in θ becomes infinite. Internal motion stops and both velocities $d\tau/dt$ (cf. Langevin twins) and $u = \sqrt{2k_B T/m} \rightarrow 0$ (the Doppler width and the black body radiation shift vanish as the thermal decoherence time increases).

To measure Boltzmann's constant several methods, particularly acoustic (propagation of sound in a gas), electrical (Johnson noise) and optical (Doppler width measurements), are presently being studied (Bordé & Himbert, 2009). They convey the hope of a low enough uncertainty (about 10^{-6}) to consider a new definition of the kelvin from Boltzmann's constant later on. In principle, such a redefinition does not face objections, and so it could be done as soon as two different methods agree at the required accuracy. The Boltzmann constant comes into play at the microscopic level through its ratio to the Planck constant and at the macroscopic level through its product by the Avogadro constant. Any future redefinition of the kelvin should take into account one of these associations, according to the future definition of the unit of mass.

10   What about the time unit? towards a totally unified system

The measurement of time is the tip top of metrology. The accuracy of atomic clocks has steadily increased by a factor of 10 every ten years, and this rate has even

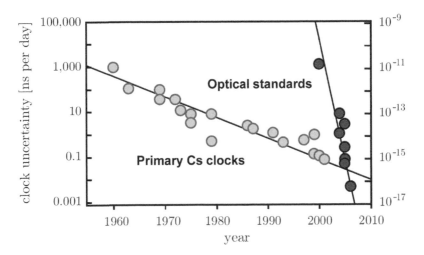

Figure 3.7 Evolution of the frequency accuracy of atomic clocks.
(from Riehle, 2015)

accelerated recently with the advent of optical clocks (Riehle, 2015; Abgrall et al., 2015; Salomon, 2015). Today their uncertainty reaches 10^{-18} (see Figure 3.7).

Thanks to this very high level of accuracy, time and frequency measurements draw up the measurement of all other quantities. This progress has its roots in the most recent atomic physics with cold atoms, and it finds everyday new applications, such as the global positioning satellite system (GPS). The teams of SYRTE at the Paris Observatory and at the Kastler-Brossel Laboratory were pioneers in the use of cold atoms to create clocks with atomic fountains. Among new revolutions, we can quote the optical clocks which, together with the frequency combs given by the femtosecond lasers (J. L. Hall and T. W. Haensch, Nobel Prize 2005), will permit to count better and faster, and there is every chance they will take the place of the microwave clocks in the future. The competition runs high between neutral atoms (in free flight or trapped in a light grating to benefit from the Lamb-Dicke effect) and trapped ions. In the end, what part will the space equipment play when it comes to compare clocks and to distribute time? In the future the use of clocks on Earth will inevitably be stopped at the level 10^{-17} by the lack of knowledge of the terrestrial gravitational potential. Then an orbital reference clock will be needed. In the future who will be the masters of time?

The future possible redefinitions of the second are an open debate. Will the second have, like the metre, a universal definition assorted with a way to put it in practice, plus secondary realizations? This would raise, just as in the case of the metre, the question of a possible variation of the fundamental constants that would modify differently the different retained transitions. The rubidium has better collisional properties than the caesium, and its hyperfine transition has been recommended by the CCTF (Consulting Committee for Time and

Frequency) as a secondary representation of the time unit. On its side, hydrogen attracts many metrological physicists who would like the definition of the time unit to be based on its transition 1s-2s. That transition was the subject of spectacular intercomparisons (at 10^{-14}) with a cold caesium fountain. A suitable combination of optical frequencies could also be used to best isolate the Rydberg constant from various corrections. The calculation of the hydrogen spectrum should be carried as far as possible, at the same time keeping in mind the considerable gap that will still separate theory from experience for a long time. Last but not least, between the Rydberg constant and the electron mass m_e we have the fine structure constant that is known only up to $0.7.10^{-9}$ so far, either by measuring the anomalous magnetic moment of the electron or more recently by atom interferometry (Cadoret et al., 2008; Bouchendira et al., 2011). Obviously, there is still a long way to formally tie the time unit to a fundamental constant, but we must be aware of the implicit link existing between the definition of the time unit and these fundamental constants. In fact, this situation is generic: owing to the permanent gap between theory and naturally reproducible phenomena, we might never be able to define units in terms of fundamental constants only. At some point we are satisfied with formulas that describe the phenomenon until we discover corrections that are too complex to be evaluated, and we have then the choice between having a simple definition from a fundamental theory or the use of a complex but very reproducible experimental procedure. This is the situation for the time unit now. In any case, let us recall that the frequency provided by an atomic clock should be corrected not only from the influence of all external fields, but also of Doppler and recoil shifts in order to yield a true atom Bohr frequency, and that such a Bohr frequency is the difference between two de Broglie-Compton frequencies. Should Planck's constant be fixed to define the unit of mass, the time unit would therefore always be defined by the difference between two masses of an atomic species. It is our choice to select either masses of elementary particles with the advantage of simplicity or masses corresponding to internal states of very complex objects far from fundamental physics, but with the possible advantage of a better reproducibility.

As a finishing touch to this quick survey of the base units and their connection with fundamental constants, let us emphasize that a new metrology in which quantum mechanics plays a more and more important part is building up.

Presently (in 2015, date of writing this article), the whole scientific community is still urged to throw light on the different choices aiming at the final decision at the CGPM of 2018.

11   A generalized framework for fundamental metrology: 5D geometry combining space-time and proper time

Beyond the choice of relevant fundamental constants, we must give coherence to the new system of units. To obtain a consistent approach to this system we must inscribe it in a unified physical framework for fundamental metrology that contains a proper description of space-time, proper time and mass, and it includes gravitation and electromagnetism as the main interactions. This goal can be reached on purely geometrical grounds as we show in Appendix 2.

This 5D scheme includes General Relativity with a 4D metric tensor $g^{\mu\nu}$ and an electromagnetic 4-potential A_μ (with μ, ν = 0, 1, 2, 3), thanks to a metric tensor $G_{\hat\mu\hat\nu}$ for 5D such that the generalized interval given by:

$$d\sigma^2 = G_{\hat\mu\hat\nu}d\hat{x}^{\hat\mu}d\hat{x}^{\hat\nu} \text{ with } \hat\mu, \hat\nu = 0, 1, 2, 3, 4 \tag{10}$$

is an invariant.

This metric tensor in five dimensions $G_{\hat\mu\hat\nu}$ is written as in Kaluza's theory (Kaluza, 1921) to include the electromagnetic gauge field potential A_μ:

$$
G_{\hat\mu\hat\nu} = \begin{pmatrix} G_{\mu\nu} & G_{\mu 4} \\ G_{4\nu} & G_{44} \end{pmatrix} = \begin{pmatrix} g_{\mu\nu} - \kappa^2 A_\mu A_\nu & -\kappa A_\mu \\ -\kappa A_\nu & -1 \end{pmatrix}
$$
$$
G^{\hat\mu\hat\nu} = \begin{pmatrix} G^{\mu\nu} & G^{\mu 4} \\ G^{4\nu} & G^{44} \end{pmatrix} = \begin{pmatrix} g^{\mu\nu} & -\kappa A^\mu \\ -\kappa A^\nu & -1+\kappa^2 A^\mu A_\mu \end{pmatrix}
\tag{11}
$$

where κ is given by the gyromagnetic ratio of the object.[8] It is such that the equation:

$$G^{\hat\mu\hat\nu}\hat{p}_{\hat\mu}\hat{p}_{\hat\nu} = 0 \tag{12}$$

with

$$\hat{p}_{\hat\mu} = (p_\mu, -mc) \tag{13}$$

and $G_{44} = -1$ is equivalent to the usual equation in 4D for a massive particle of mass m and charge q:

$$g^{\mu\nu}(p_\mu - qA_\mu)(p_\nu - qA_\nu) = m^2 c^2 \tag{14}$$

These last equations give directly the Klein-Gordon equation in 5D for the field ϕ:

$$\hat\Box\varphi = G^{\hat\mu\hat\nu}\hat\nabla_{\hat\mu}\hat\nabla_{\hat\nu}\varphi = 0 \tag{15}$$

where the connection between mechanical quantities and quantum-mechanical operators is made as usual through Planck's constant. This equation is analogous to the wave equation for massless particles in 4D.

The phase of this field is given by the 5D-superaction $\hat{S}$ in units of $\hbar$ and satisfies the Hamilton-Jacobi equation:

$$G^{\hat\mu\hat\nu}\partial_{\hat\mu}\hat{S}\partial_{\hat\nu}\hat{S} = 0 \tag{16}$$

With this geometrical picture we have gathered all quantities concerned by the main base units: space-time, proper time, mass, gravito-inertial and electromagnetic fields in a phase without dimension. Any measurement can then be reduced to a phase measurement through a suitable interferometry experiment, since all base quantities enter the expression of a phase through a comparison with reference quantities of the same nature. This universal link is obtained by fixing Planck's constant. Mass and proper time are entangled concepts which correspond to conjugate variables in classical mechanics and to non-commuting operators in quantum mechanics in complete analogy with momentum and position operators. The photon box of the Einstein-Bohr controversy is a direct illustration of this quantum behaviour and of the non-commuting character of proper time and mass operators:

$$[c\tau_{op}, m_{op}c] = i\hbar \tag{17}$$

Their respective units thus require a joint definition in which the unit of mass is defined from the mass difference of the two levels involved in the definition of the unit of time. A compatible *mise en pratique* requires us to associate a quantum clock with a macroscopic mass through a phase measurement, either by atom interferometry and atom counting or in the watt balance. The Avogadro constant is then obtained directly from the measurement of the de Broglie-Compton frequency of the carbon atom in a recoil experiment.

The proper time acquires a status in quantum mechanics, and we may now describe the quantum theory of atomic clocks in general relativity from their internal properties, since the phase of atom waves can be corrected from general relativistic effects such as the gravitational red shift (Bordé, 2002).

Finally, temperature and time can be combined in a complex time variable in the theory of clocks. This accounts for thermal decoherence through the Doppler shift in atom interferometers. A generalized line shape for the usual Doppler broadening can be derived accordingly (Bordé, 2009).

12   Conclusion

Most base quantities of metrology, length, time, mass, electrical quantities and temperature are ultimately measured by optical or matter-wave interferometers. Optics and quantum mechanics play a central role in the description of these devices. Consequently, future fundamental metrology will deal essentially with phase measurements i.e. invariant numbers. One should also emphasize the non-commuting character of quantities like mass and proper time, which is a reason why Planck's constant has such a special place in the system of units. Base quantities should be quantum observables. Some appear as base quantities with their conjugate partner (e.g. mass and proper time), others do not (e.g. position coordinate and momentum). The quantum-mechanical link between conjugate quantities does not allow any more to leave Planck's constant out of the system of units, which would be the case if the mass unit continued to be defined by the mass of an object, whether macroscopic ($\mathfrak{K}$) or microscopic (atom or electron). We have seen that a natural choice was to couple the definitions for mass and time units. Heisenberg's uncertainty relations will apply in the quantum limit.[9] Measurement theory becomes essential to explore the limits of the new quantum metrology.

A natural 5D theoretical framework for the redefinition of the SI is completely provided by the connection between pure geometry, metric tensor and metrology, that we have outlined. In this way, a clear separation has been made between proper time (observable!) and time coordinate (not observable!) as distinct quantities sharing the same unit. The role of the electromagnetic field is to couple space-time and proper time coordinates through the corresponding off diagonal components of the metric tensor. The 5D action gathers all phenomena and constants of interest for a fully relativistic quantum metrology in an invariant phase through Planck's constant and this includes the dephasing arising from gravito-inertial fields (e.g. the Sagnac effect or the effect of gravitational waves) as well as those of electromagnetic origin (such as the Aharonov-Bohm or the Aharonov-Casher effect).

Reducing the theory of measurements to the determination of quantum phases was our primary objective, and this chapter is a first attempt to go in this direction and to unify all aspects of modern quantum metrology. The perspective that we have adopted, incorporates naturally all relevant fundamental constants in a logical scheme with obvious constraints of economy, aesthetics and rigour. The final aim is, of course, to adopt a system of units free of arbitrary and artificial features, in harmony with contemporary physics.

Acknowledgements

This chapter, written in 2015 and submitted on October 19, 2015, is adapted from an article in "La lettre de l'Académie des sciences" n°20 (2007) which was kindly translated to English by Mrs. Marie-Claire Céreuil. The author is also grateful to a number of colleagues for constructive comments, especially Dr. Olivier Darrigol, Dr. Nadine de Courtenay, Dr. Franck Pereira and Pr. Marc Himbert.

Appendix 1
What framework for relativistic quantum metrology?[10]

This framework is naturally the one imposed by the two great physical theories of the twentieth century: relativity and quantum mechanics. These two major theories themselves have given birth to quantum field theory, which incorporates all their essential aspects and adds those associated with quantum statistics. The quantum theory of fields allows a unified treatment of fundamental interactions, especially, of electroweak and strong interactions within the standard model. General Relativity is a classical theory; hence gravitation remains apart and is reintegrated into the quantum world only in the recent theories of strings. We do not wish to go that far, and we will keep to quantum electrodynamics and to the classical gravitation field. Such a framework is sufficient to build a modern metrology, taking into account an emerging quantum metrology. Of course, quantum physics has been operating for a long time at the atomic level, for example in atomic clocks, but now it also fills the gap between this atomic world and the macroscopic world, thanks to the phenomena of quantum interferences whether concerning photons, electrons, Cooper pairs or more recently atoms in atom interferometers (Bordé, 1989).

The main point is to distinguish between a "kinematical" framework associated with fundamental constants having a dimension, such as c, $\hbar$, k_B, and a "dynamical" framework where the interactions are described by coupling constants without dimension. The former framework relies on the Statistical Relativistic Quantum Mechanics of free particles, and the latter on the quantum field theory of interactions.

Two possible goals can be pursued:

i redefine each unit in terms of a fundamental constant with the same dimension, e.g. mass in terms of the mass of an elementary particle; or
ii reduce the number of independent units by fixing a fundamental constant having the proper dimension for this reduction, e.g. fixing c to connect space and time units or – to connect mass and time units.

The existence of fundamental constants with a dimension is often an indication that we are referring to the same thing with two different names. We recognize this identity as our understanding of the world gets deeper. We should then apply an economy principle (Occam's razor) to our unit system to take this into account and to display this connection.

When abandoning a unit for the sake of another, the first condition (C1) is thus to recognize an equivalence between the quantities measured with these units (e.g. equivalence between heat and mechanical energy and between mass and energy), or a symmetry of nature that connects these quantities in an operation of symmetry (for example a rotation transforming the space coordinates into one another or of a Lorentz transformation mixing the space and time coordinates).

A second condition (C2) is that a realistic and mature technology of measurement is to be found. For example, notwithstanding the equivalence between mass and energy, in practice the kilogram standard will not be defined by an energy of annihilation, but on the other hand, thanks to the watt balance, it can be tied to its Compton frequency $M_R c^2/h$ by measurements of time and frequency.

A third condition (C3) is connected to the confidence felt for the understanding and the modelization of the phenomenon used to create the link between quantities. For instance, the exact measurement of distances by optical interferometry is never questioned because we believe that we know everything, and in any case, that we know how to calculate everything concerning the propagation of light. That is the reason why redefining the metre ultimately took place without much problem. On the other hand, measuring differences of potential by the Josephson effect or electrical resistances by the quantum Hall effect still needs support because despite a 10^{-9} confirmation of their reproducibility and a good understanding of the universal topological character of these phenomena, some people still feel uncertain as to whether all possible small parasitical effects have been dealt with. For a physical phenomenon to be used to measure a quantity properly, it must be directly related to our knowledge of the whole underlying physics. In order to switch to a new definition, this psychological barrier must be overcome, and we must have complete faith in our total understanding of the essentials of the phenomenon. Therefore, through a number of experiments as varied as possible, we must make sure that the measurement results are consistent up to a certain level of accuracy which will be that of the *mise en pratique*, and we must convince ourselves that no effect has been neglected at that level. If all of these conditions are fulfilled, the measured constant that linked the units for the two quantities will be fixed, e.g. the mechanical equivalent of the calorie or the speed of light.

Appendix 2

The status of mass in classical relativistic mechanics from 4 to 5 dimensions

In special relativity, the total energy E and the momentum components p^1, p^2, p^3 of a particle transform as the contravariant components of a four-vector

$$p^\mu = (p^0, p^1, p^2, p^3) = (E/c, \vec{p})$$ (18)

and the covariant components are given by:

$$p_\mu = g_{\mu\nu} p^\nu$$ (19)

where $g_{\mu\nu}$ is the metric tensor. In Minkowski space of signature $(+, -, -, -)$:

$$p_\mu = (p_0, p_1, p_2, p_3) = (E/c, -p^1, -p^2, -p^3)$$ (20)

These components are conserved quantities when the considered system is invariant under corresponding space-time translations. They will become the generators of space-time translations in the quantum theory. For massive particles of rest mass m, they are connected by the following energy-momentum relation:

$$E^2 = p^2 c^2 + m^2 c^4$$ (21)

or, in manifestly covariant form:

$$p^\mu p_\mu - m^2 c^2 = 0$$ (22)

This equation cannot be considered as a definition of mass since the origin of mass is not in the external motion, but rather in an internal motion. It simply relates two relativistic invariants and gives a relativistic expression for the total energy. Thus, mass appears as an additional momentum component, mc, corresponding to internal degrees of freedom of the object, which adds up quadratically with external components of the momentum to yield the total energy squared (Pythagorean theorem). In the reference frame in which $p = 0$, the squared mass term is responsible for the total energy, and mass can thus be seen as stored internal energy, just like kinetic energy is a form of external energy. Even when this internal energy is purely kinetic, e.g. in the case of a photon in a box, it appears as pure mass m^* for the global system (i.e. the box). This new mass is the relativistic mass of the stored particle:

$$m^*c^2 = \sqrt{p^2c^2 + m^2c^4} \tag{23}$$

The concept of relativistic mass has been criticized in the past, but it becomes relevant for embedded systems. We may have a hierarchy of composed objects (e.g. nuclei, atoms, molecules, atomic clock), and at each level the mass m^* of the larger object is given by the sum of energies p^0 of the inner particles. It transforms as p^0 with the internal coordinates and is a scalar with respect to the upper level coordinates.

Mass is conserved when the system under consideration is invariant in a proper time translation and will become the generator of such translations in the quantum theory. In the case of atoms, the internal degrees of freedom give rise to a mass which varies with the internal excitation. For example, in the presence of an electromagnetic field inducing transitions between internal energy levels, the mass of atoms becomes time-dependent (Rabi oscillations). It is thus necessary to enlarge the usual framework of dynamics to introduce this new dynamical variable as a fifth component of the energy-momentum vector.

Equation (22) can be written with a five dimensional notation:

$$G^{\hat\mu\hat\nu}\hat p_{\hat\mu}\hat p_{\hat\nu} = 0 \text{ with } \hat\mu, \hat\nu = 0, 1, 2, 3, 4 \tag{24}$$

where $\hat p_{\hat\mu} = (p_\mu, p_4 = -mc)$; $G^{\mu\nu} = g^{\mu\nu}$; $G^{\hat\mu 4} = G^{4\hat\nu} = 0$; $G^{44} = G_{44} - 1$

This leads us to consider also the picture in the coordinate space and its extension to five dimensions. As in the previous case, we have a four-vector representing the space-time position of a particle:

$$x^\mu = (ct, x, y, z)$$

and in view of the extension to general relativity:

$$dx^\mu = (cdt, dx, dy, dz) = (dx^0, dx^1, dx^2, dx^3) \tag{25}$$

The relativistic invariant is, in this case, the elementary interval ds, also expressed with the proper time τ of the particle:

$$ds^2 = dx^\mu dx_\mu = c^2 dt^2 - d\vec{x}^2 = c^2 d\tau^2 \tag{26}$$

which is, as that was already the case for mass, equal to zero for light

$$ds^2 = 0 \tag{27}$$

and this defines the usual light cone in space-time.

For massive particles, proper time and interval are non-zero, and equation (26) defines again a hyperboloid. As in the energy-momentum picture, we may enlarge our space-time with the additional dimension $s = c\tau$

$$d\hat{x}^{\hat{\mu}} = (cdt, dx, dy, dz, cd\tau) = (dx^0, dx^1, dx^2, dx^3, dx^4) \tag{28}$$

and introduce a generalized light cone for massive particles:[11]

$$d\sigma^2 = G_{\hat{\mu}\hat{\nu}} d\hat{x}^{\hat{\mu}} d\hat{x}^{\hat{\nu}} = c^2 dt^2 - d\overline{x}^2 - c^2 d\tau^2 = 0 \tag{29}$$

As pointed out in the case of mass, proper time is not defined by this equation from other coordinates, but it is rather a true evolution parameter representative of the internal evolution of the object. It coincides only numerically with the time coordinate in the frame of the object through the relation:

$$cd\tau = \sqrt{G_{00}} dx^0 \tag{30}$$

Finally, if we combine momenta and coordinates to form a mixed scalar product, we obtain a new relativistic invariant that is the differential of the action. In 4D:

$$dS = -p_\mu dx^\mu \tag{31}$$

and in 5D we shall therefore introduce the superaction:

$$\hat{S} = -\int \hat{p}_{\hat{\mu}} d\hat{x}^{\hat{\mu}} \tag{32}$$

equivalent to

$$\hat{p}_{\hat{\mu}} = -\frac{\partial \hat{S}}{\partial \hat{x}^{\hat{\mu}}} \text{ with } \hat{\mu} = 0, 1, 2, 3, 4 \tag{33}$$

If this is substituted in

$$G^{\hat{\mu}\hat{\nu}} \hat{p}_{\hat{\mu}} \hat{p}_{\hat{\nu}} = 0 \tag{34}$$

we obtain the Hamilton-Jacobi equation in 5D

$$G^{\hat{\mu}\hat{\nu}} \partial_{\hat{\mu}} \hat{S} \partial_{\hat{\nu}} \hat{S} = 0 \tag{35}$$

which has the same form as the eikonal equation for light in 4D. It is already this striking analogy that pushed Louis de Broglie to identify action and the phase of a matter wave in the 4D case. We shall follow the same track for a quantum approach in our 5D case.

What is the link between the three previous invariants given? As in optics, the direction of propagation of a particle is determined by the momentum vector tangent to the trajectory. The 5D momentum can therefore be written in the form:

$$\hat{p}^{\hat{\mu}} = d\hat{x}^{\hat{\mu}} / d\lambda \tag{36}$$

where λ is an affine parameter varying along the ray. This is consistent with the invariance of these quantities for uniform motion. In 4D the canonical 4-momentum is:

$$p_{\mu} = mc \frac{g_{\mu\nu} dx^{\nu}}{\sqrt{g_{\mu\nu} dx^{\mu} dx^{\nu}}} = mc g_{\mu\nu} u^{\nu} \tag{37}$$

where $u^{\nu} = dx^{\nu}/d\tau$ is the normalized 4-velocity with $d\tau = \sqrt{g_{\mu\nu} dx^{\mu} dx^{\nu}}$ given by (26). We observe that $d\lambda$ can always be written as the ratio of a time to a mass:

$$d\lambda = \frac{d\tau}{m} = \frac{dt}{m^*} = \frac{d\theta}{M} = ... \tag{38}$$

where τ is the proper time of individual particles (e.g. atoms in a clock or in a molecule), t is the time coordinate of the composed object (clock, interferometer or molecule) and θ its proper time; m, m^*, M are respectively the mass, the relativistic mass of individual particles and their contribution to the scalar mass of the device or composed object.

From

$$d\sigma^2 = G_{\hat{\mu}\hat{\nu}} d\hat{x}^{\hat{\mu}} d\hat{x}^{\hat{\nu}} = 0 \tag{39}$$

we infer that in 5D

$$d\hat{S} = 0 \tag{40}$$

whereas in 4D

$$dS = -p_{\mu} dx^{\mu} = -mc^2 d\tau \tag{41}$$

As a consequence, the quantum-mechanical phase also cancels along the classical trajectory in 5D. The particle is naturally associated with the position where all phases cancel to generate a constructive interference.

The previous 5D scheme can be extended to General Relativity with a 4D metric tensor $g^{\mu\nu}$ and an electromagnetic 4-potential A_{μ} (Bordé, 2014) with the metric tensor given in the main text.

Notes

1 We should carefully distinguish two different meanings of time: on the one hand, time and position mix as coordinates, and this refers to the concept of time coordinate for an event in space-time, which is only one component of a 4-vector; on the other hand, time is the evolution parameter of a composite system, and this refers to the proper time of this system, and it is a Lorentz scalar.

2 There is presently no physical clock at the de Broglie-Compton frequency, although it appears quite possible in the future through stimulated absorption and emission of photon pairs in the creation/annihilation process of electron-positron pairs.

3 A well-defined phase assumes that the object should be in an eigenstate of its internal Hamiltonian. In the case of a collection of atoms, this could be realized only with a large Bose-Einstein condensate.

4 *Comité consultatif pour la masse et les grandeurs apparentées.*

5 *Comité international des poids et mesures.*

6 Committee on Data for Science and Technology.

7 See, for example, the theory of linear absorption of light by gases and its application to the determination of Boltzmann's constant (Bordé, 2009).

8 In the case of the electron: $\kappa = e/m_e c$, which can also be written as

$$\frac{1}{c}\sqrt{\frac{\alpha}{\alpha_G}}\sqrt{4\pi\varepsilon_0 G}$$

where we have introduced the dilaton field $\sqrt{\alpha/\alpha_G}$ to make the connection with Kaluza's theory (Kaluza, 1921).

9 One should also keep in mind the uncertainty relation between phase and number of entities, i.e. between action and quantity of matter.

10 The following discussion is reproduced from references Bordé, 2004, 2005.

11 In this picture, anti-particles have a negative mass and propagate backwards on the fifth axis as first pointed out by Feynman. Still, their relativistic mass m^* is positive, and hence they follow the same trajectories as particles in gravitational fields as one can check from the equations of motion (Bordé, 2014).

References

Abgrall, M., Chupin, B., De Sarlo, L., Guéna J., Laurent, P., Le Coq, Y., . . . Bize, S. (2015). Atomic fountains and optical clocks at SYRTE: Status and perspectives. *Comptes Rendus Physique, 16*(5), 461–470.

Andreas, B., et al. (2011). Counting the atoms in a (28)Si crystal for a new kilogram definition. *Metrologia, 48*, S1–S13.

Barger, R. L., & Hall, J. L. (1969). Pressure shift and broadening of methane line at 3.39 μm studied by laser-saturated molecular absorption. *Physical Review Letters, 22*, 4–8.

Berman, P. (Ed.). (1997). *Atom interferometry*. Cambridge, MA: Academic Press.

Bordé, Ch. J. (1970). Spectroscopie d'absorption saturée de diverses molécules au moyen des lasers 'a gaz carbonique et 'a protoxyde d'azote. *Comptes Rendus de l'Académie des Sciences Paris, 271B*, 371–374.

Bordé, Ch. J. (1989). Atomic interferometry with internal state labelling, *Physics Letters, A140*, 10–12.

Bordé, Ch. J. (1997). *Matter-wave interferometers: A synthetic approach*. Cambridge, MA: Academic Press.

Bordé, Ch. J. (2002). Atomic clocks and inertial sensors. *Metrologia, 39*(5), 435–463.

Bordé, Ch. J. (2004). Métrologie fondamentale: unités de base et constantes fondamentales. *C.R. Physique*, 5, 813–820.

Bordé, Ch. J. (2005). Base units of the SI, fundamental constants and modern quantum physics. *Philosophical Transactions of the Royal Society A: Mathematical, Physical and Engineering Sciences*, 363, 2177–2202, 2182.

Bordé, Ch. J. (2009). On the theory of linear absorption line shapes in gases. *C.R. Physique*, 10, 866–882.

Bordé, Ch. J. (2014). Atom interferometry using internal excitation: Foundations and recent theory. In *International School of Physics "Enrico Fermi" COURSE CLXXXVIII Atom Interferometry* (pp. 143–170). Bologna, Italy: Società Italiana di Fisica.

Bordé, Ch. J., & Himbert, M. E. (2009). Experimental determination of Boltzmann's constant. Special issue of the *Comptes Rendus de l'Académie des Sciences*, 10, 813.

Bouchendira, R., Cladé, P., Guellati-Khélifa, S., Nez, F., & Biraben, F. (2011). New determination of the fine structure constant and test of the quantum electrodynamics. *PRL*, 106, 080801.

Broglie, L. de (1923). Ondes et quanta. *Comptes Rendus de l'Académie des Sciences*, 2, 507.

Cadoret, M., de Mirandes, E., Cladé, P., Guellati-Khélifa, S., Schwob, C., Nez, F., Julien, L., & Biraben, F. (2008). Combination of Bloch Oscillations with a Ramsey Bordé Interferometer: New determination of the fine structure constant. *PRL*, 101, 230801.

Hall, J. L., & Bordé, Ch. J. (1974). Direct resolution of the recoil doublets using saturated absorption techniques. *Bulletin of the American Physical Society*, 19, 1196.

Hall, J. L., Bordé, Ch. J., & Uehara, K. (1976). Direct optical resolution of the recoil effect using saturated absorption spectroscopy. *Physical Review Letters*, 37, 1339–1342.

Kaluza, T. (1921). Zum Unit atsproblem der Physik. *Sitzungsberichte der Preussischen Akademie der Wissenschaften Physikalisch-Mathematische Klasse*, K1, 966.

Kibble, B. P. (1975). A measurement of the gyromagnetic ratio of the proton by the strong field method. In J. H. Sanders and A. H. Wapstra (Eds.), *Atomic an fundamental constants* (Vol. 5, pp. 545–551). New York, NY: Plenum Press.

Parker, W. H., & Simmonds, M. B. (1971). Measurement of h/m_e using rotating superconductors. In *M. B. precision measurement and fundamental constants* (pp. 243–247). Washington D.C: NBS Special Publication.

Piquemal, F., Bounouh, A., Devoille, L., Feltin, N., Thevenot, O., & Trapon, G. (2004). Fundamental electrical standards and the quantum metrological triangle. *Comptes Rendus Physique*, 5(8), 857–879.

Riehle, F. (2015). Towards a redefinition of the second based on optical atomic clocks. *Comptes Rendus Physique*, 16(5), 506–515.

Salomon, C. (2015). The measurement of time. Special issue of the *Comptes Rendus de l'Académie des Sciences*, 16.

Taylor, B. N., & Mohr, P. J. (1999). On the redefinition of the kilogram. *Metrologia*, 36, 63–64.

Taylor, B. N., & Mohr, P. J. (2001). The role of fundamental constants in the International System of Units (SI): Present and future. *IEEE Transactions on Instrumentation and Measurement*, 50, 563–567.

Vocke Jr., R. D., Rabb, S. A., & Turk, G. C. (2014). Absolute silicon molar mass measurements, the Avogadro constant and the redefinition of the kilogram. *Metrologia*, 51, 361–375.

Wignall, J. W. G. (1992). Proposal for an absolute, atomic definition of mass. *Physical Review Letters*, *68*, 5–8.

Young, B., Kasevich, M., & Chu, S. (1997). Precision atom interferometry with light pulses. In *Atom interferometry* (pp. 363–406). Cambridge, MA: Academic Press.

Zimmerman, J. E., & Mercereau, J. E. (1965). Compton wavelength of superconducting electrons. *Physical Review Letters*, *14*, 887–888.

4 Strategies for the definition of a system of units

Alessandro Giordani and Luca Mari

1 Introduction

Quantity units[1] give an empirical reference to numerical values in mathematical equations that represent physical laws. Units have then the role of relating theory and experiment, and a system of units is a critical component of what metrology has to offer to empirical sciences. We analyse here the strategic problem of the optimal criteria – in a sense to be specified – to define a system of units, where such a system is specified when (a) a system of quantities and (b) a unit for each quantity in the system are defined. For the definition of a system of units, the previous knowledge of a system of quantities is then assumed, as typically provided by physics or a subfield of it. The focus is on the strategies according to which quantity units can be defined.[2] Hence what follows applies not only to the International System of Units (SI), in its current version (BIPM, 2006) and its possible future revision (BIPM, 2014), but also and more generally to the definition of any system of units, and it is aimed at discussing and providing a strategic perspective to such definition. After two sections devoted to the introduction of the basic relevant concepts and the main decision criteria for the selection of a system of units, in Section 4 the general features of a strategy for the definition of such a system are discussed and exemplified by means of a simple system. Constant-based definitions are then particularly analysed in Section 5, thus leading to the identification, in Section 6, of a structurally best strategy and to a possible improvement of it.

2 Basic concepts

The conceptual framework for the analysis that we are going to develop is based on a set of fundamental categories and three groups of more specific concepts related to (a) systems of quantities, (b) quantity units and systems of units and (c) realization of the definition of quantity units. In the following subsections an essential account of such concepts is provided.[3]

2.1   Categories

The following fundamental categories of entities are assumed:

i   object;
ii   quantity;
iii   kind of object;
iv   kind of quantity.

The entities falling under these categories are related as in Figure 4.1.

We will denote with $s : S$ the fact that s is an entity of kind S, where then $Ext(S) = \{s \mid s : S\}$ is the extension of S.

The previous concepts can be exemplified as follows. A given electron is an object of a given kind, say <electron>: it is therefore an instance of that kind of object. The electric charge of a given electron is a quantity, to be intended as an individual quantity, of a given kind, say <electric charge>, which is a general quantity: it is therefore an instance of that kind of quantity.[4] A given electron is characterized by an individual electric charge, and therefore that electron has that electric charge. By inheritance, any electron is characterized by the kind of quantity <electric charge>, i.e., it has an electric charge. In turn, the kind of object <electron> is characterized by the kind of quantity <electric charge>. Finally, the model of a kind of object might specify that there exists a kind of quantity such that all objects of that kind have the same individual quantity. It is the case of electrons and electric charge: according to the current model each electron is characterized by the same electric charge, so that the kind of object <electron> is characterized not only by the kind of quantity <electric charge> but also by a given individual quantity. Hence the kind of object <electron> is characterized by that individual electric charge. We will say that an object which is characterized by a certain individual quantity is the support of that quantity, and a kind of object which is characterized by an individual quantity according to a certain model is the ideal support of that quantity.

The relations depicted in the diagram can be considered from the complementary perspectives of idealization ("bottom-up") and realization ("top-down").

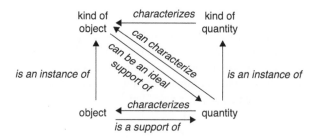

Figure 4.1 Fundamental categories of entities and the relations holding between them.

For example, the observation of the indistinguishability of the electric charge of distinct electrons may lead to a model of electron, as a kind of object, such that all objects of the kind <electron> are characterized by the same electric charge. In a complementary way, the assumption of the model implies that an object is of the kind <electron> if its electric charge equals the one predicted by the model.

2.2 Concepts related to systems of quantities

The main concepts assumed related to systems of quantities[5] are:

i related kinds of quantity;
ii system of quantities;
iii dimension of a quantity.

The kinds of quantity that characterize a given kind of object can be mutually related. An essential part of the development of experimental sciences consists indeed of the discovery of empirical laws about the relations between kinds of quantity and their mathematical representation. A system of quantities is a set of mutually related kinds of quantity, in which a conventional distinction is introduced between base kinds of quantity and derived kinds of quantity, in such a way that the latter can be functionally derived from the former through relations in the system.

More precisely, let $\mathbf{Q} = \{Q_i\}_{1 \leq i \leq N}$ be a set of kinds of quantity and $\mathbf{R} = \{R_k\}_{1 \leq k \leq K}$ be a set of relations between elements of $\mathbf{Q}$, as specified according to a set of agreed empirical laws. Then, a basis for $\mathbf{Q}$ is a set $\mathbf{B} \subseteq \mathbf{Q}$, such that every kind of quantity in $\mathbf{Q}$ can be represented in terms of elements in $\mathbf{B}$ given the relations in $\mathbf{R}$.[6] In the case of the set of physical quantities of interest for the SI, this general characterization can be specified: every kind of quantity Q_i in $\mathbf{Q}$ can be represented as a product of powers of elements of the basis $\mathbf{B} = \{B_j\}_{1 \leq j \leq J}$, as

$$Q_i = \prod_{1 \leq j \leq J} B_j^{n_{ij}} \tag{1}$$

where $n_{ij} = \delta_{ij}$, if $Q_i \in \mathbf{B}$.[7]

The functional dependence of a kind of quantity on the base quantities of a system is the dimension of that quantity in the given system. Hence, through equation (1), each kind of quantity in a system is provided with a dimension, and base kinds of quantity are identifiable as the ones that coincide with their dimension. While the relations in a system of quantities are invariant with respect to the choice of base quantities, the dimension of quantity does depend on this choice. For example, suppose $\mathbf{Q} = \{T = \text{<time>}, L = \text{<length>}, V = \text{<speed>}\}$.[8] Then, the relation $L = TV$ is unaffected by the choice of base quantities. Still, if the set of base quantities $\mathbf{B} = \{T, L\}$ is chosen, then dim $L = L$, whereas if $\mathbf{B} = \{T, V\}$, then dim $L = TV$. Finally, it is assumed that the dimension of an individual quantity is the dimension of the corresponding kind.

2.3   *Concepts related to quantity units and systems of units*

The main concepts assumed related to quantity units and systems of units are:

  i   quantity unit;
  ii   definition of a quantity unit;
  iii   system of quantity units;
  iv   reference quantity in the definition of a unit;
  v   reference object in the definition of a unit;
  vi   reference kind of object in the definition of a unit;
  vii   kinds of quantity defining a unit;
  viii   independent definition of a unit;
  ix   dependent definition of a unit.

A unit for a given kind of quantity is an instance of that kind, chosen by convention and identified through a definition. In case of a ratio kind of quantity Q, the choice of a unit $[Q]$ allows us to represent any individual quantity $q : Q$ as

$$q = \{q\}[Q] \tag{2}$$

where $\{q\}$, called the numerical quantity value, is then the ratio between q and the unit $[Q]$, two individual quantities that are instances of the same kind.[9] A system of units is a set of quantity units related in such a way that the relations between the units reflect the relations between the corresponding kinds of quantity. A system of quantities is said to be coherent when the relations between the units identically reflect the relations between the corresponding kinds of quantity:

$$[Q_i] = \Pi_{1 \leq j \leq J} [B_j]^{n_{ij}} \tag{3}$$

where $[B_j]$ is a base unit, i.e., the unit corresponding to a base kind of quantity.

Since a unit $[Q]$ is an instance of a given kind of quantity Q, the structurally simplest option is to define $[Q]$ by just selecting an instance q of Q:

$$[Q] := q \tag{4}$$

A second option is to define the unit by reference to a quantity of the same kind of the unit and related to the unit through a numerical coefficient k:

$$[Q] := k \, q \tag{5}$$

A reference quantity in the definition of a unit is then an individual quantity that is specified in the definition of a unit. A reference object in the definition of a unit is a support of a reference quantity. A reference kind of object in the definition of a unit is an ideal support of a reference quantity, where the ideal character stems from the assumption that each object of that kind is characterized by the same individual quantity.

A third option to define a unit [Q] is possible if its kind Q belongs to a system of quantities, so that the relations in the system can be exploited in the definition and the unit is defined as a function f of quantities $q_1, q_2, \ldots, q_n$ of kinds $Q_1, Q_2, \ldots, Q_n, n \geq 1$, different from Q:

$$[Q] := f(q_1, q_2, \ldots, q_n) \tag{6}$$

The definition of a unit as in equation (4) or equation (5), that does not refer to the relations in the system of quantities, is independent, and dependent is instead a definition, as in equation (6), which does refer to such relations. For example, a definition of length unit based on the relation of length with time and velocity is a dependent definition.

2.4 Concepts related to the realization of the definition of quantity units

The main concepts assumed related to the realization of the definition of quantity units are:

i   quantity realizing the definition of a unit;
ii object realizing the definition of a unit;
iii calibration;
iv metrological traceability chain;
v   calibration uncertainty.

A measurement process is based on a measurement procedure that specifies a way for comparing the measurand and the unit. This comparison implies the availability of a quantity that realizes the definition of the unit through the realization of a support of that quantity. The object realizing the definition of a unit is called a measurement standard. The comparison with a unit is then generally obtained by realizing a chain of measurement standards, from a primary one to the working standard used in measurement and calibrated against the primary standard through a metrological traceability chain. In the processes of calibration that yields the chain, a calibration uncertainty is produced that generally increases in each step of the chain from the primary standard to the standard involved in the measuring system.

3   Basic criteria

The main operative target of the definition of a system of units is to provide a basis for measurement, for which quantity units are indeed essential.[10] Measurement is critical for the solution of practical problems, by providing information that is expected to be objective and intersubjective, and therefore invariant with respect to conditions and agents: similar measurement processes should produce compatible results, within the stated uncertainty, when the measurand is the same, thus requiring a shared definition of the quantity unit and compatible realizations of it. The constraint of invariance with respect to conditions and agents leads then to the following criteria of adequacy of definition of units.

3.1 Invariance with respect to conditions and agents

The same practical problem may arise in different conditions or to different agents in the same or different times, and a measurement process supporting the solution of that problem is required to produce results that do not depend on conditions or agents. This implies, in particular, that the realization of the definition of the quantity unit remains the same even if conditions or agents change. Hence, such a definition must guarantee (a) the stability of the realizations, i.e., their invariance with respect to variations of space-time regions, and (b) the intersubjectivity of the realizations, i.e., their invariance with respect to variations of agents. Hence, in the ideal case:[11]

(**C1**) the realization of a definition should produce the same result in each time;
(**C2**) the realization of a definition should produce the same result in each condition;
(**C3**) the realization of a definition should produce the same result for each agent;
(**C4**) the realization of a definition should produce a result that is operatively accessible to each agent;
(**C5**) the realization uncertainty in each time should be minimum; and
(**C6**) the realization uncertainty in each condition should be minimum,

where of course in operative situations these criteria can be met within a given uncertainty.

Criteria (**C1**)–(**C3**) impose the stability of the product of the realization. The criterion (**C3**) does not imply that the procedure of realization be the same for each agent. Rather, it allows for different procedures to be performed if they produce the same results, so that each agent may choose the procedure that is better accessible, (**C4**), provided that its realization uncertainty is minimum (**C5**), (**C6**). Consequently, a definition admitting multiple procedures of realization fulfilling (**C1**)–(**C6**) is preferable.

3.2 Universality, continuity, uncertainty attribution

The criterion (**C4**) emphasizes a condition of universality in the realization of the definition of a unit. On the other hand, there is a trade-off between theoretical and operative universality here:

> There is no overall benefit in developing a highly accurate definition that can only be realized practically with a much-increased uncertainty, or through unfeasibly complex or expensive experimentation.
>
> (Milton, Williams, & Bennett, 2007, p. 357)

Furthermore, the social role of systems of units suggests a criterion of continuity: the definition of a quantity unit should guarantee the continuity with the previous definition, so as to preserve the available knowledge base and the operative know-how and implementations, and so that all calibrations and measurements performed according to the previous definition remain valid, within acceptable limits, also according to the new definition. Such a continuity is then related both to the structure of the system, in particular the selection of base units, and to the specification of both the base units and the reference constants. Of course, *ceteris paribus*, continuity is appropriate. Finally, the decision on what quantities are modelled as uncertain according to the definition of a unit should also be taken into account, in particular whether fundamental constants have a definitional role such that they are defined with no uncertainty or their values derive from independent definitions and then include a non-null uncertainty. Of course, *ceteris paribus*, uncertainty attribution such that universal constants are not uncertain is appropriate.

None of these criteria are absolute, and the historical development of systems of units might be studied in terms of their partial fulfillments, or even violations. A rational change should be developed as a multi-criteria decision-making process that maximizes the benefits and minimizes the problems.

4 General features of a strategy for the definition of a system of units

The criteria we have mentioned ground the framework in which a strategy can be chosen for the definition of a system of units, based on the decision on alternative options as for unit definition. In this section we discuss the general features of the framework, applicable in particular to the SI (BIPM, 2006) and presented here in reference to a subsystem of the SI that, while including only three units, is sufficiently general to exemplify the whole structure of the SI itself.

Let us start by setting the terminology. In a given theoretical context we will call:

i   *fundamental object* an object that in the context is assumed to be stable relatively to one or more given quantities in given conditions; and
ii *fundamental quantity* a quantity of a fundamental object that is stable in the given conditions.

Furthermore, again in reference to a given theoretical context we will call:

i   *absolute fundamental object* a fundamental object whose stability is assumed independent of conditions; and
ii *absolute fundamental quantity* a fundamental quantity of an absolute fundamental object,

where the adjective "absolute", as applied to both objects and quantities, emphasizes the independence of conditions.

For example, a metallic sphere might be assumed to be stable with respect to mass and diameter in given conditions (no mechanical interactions with other objects, constant temperature, etc.): the sphere is then a fundamental object, and its mass and diameter are fundamental quantities. A helium atom might be assumed to be stable with respect to mass and diameter in all conditions in which it exists: the atom is then an absolute fundamental object, and its mass and diameter are absolute fundamental quantities.

According to the stability criteria mentioned in the previous section, a quantity unit is expected to be a fundamental quantity, and possibly an absolute fundamental quantity.

4.1   Types of definitions

The definition of a system of units depends on a number of choices, related to:

i   the dependence on base quantities;
ii   the introduction of scaling factors;
iii   the type of reference entities.

Systems of units which are dependent on (independent of) the introduction of base quantities will be called based (non-based) systems. A definition involving scale factors will be called a scaled definition, and systems of units which do (do not) include scaled definitions will be called scaled (non-scaled) systems. The principled independence of these choices generates four options: scaled based systems, non-scaled based systems, etc.

As to the choice of the type of reference quantities, three options are possible:

(T1) quantities of objects, thus operating as prototypes;
(T2) quantities of kinds of object; and
(T3) fundamental constant quantities, whose supports are kinds of interactions.

Furthermore, for all types the reference to more than one entity is possible, corresponding to the case of a dependent definition.

In case of based systems, while base quantities are mutually independent in the system, in the sense that the dimension of each base quantity cannot be obtained as a product of powers of dimensions of other base quantities, base units do not need to be all independent in their definition, and the definition of one or more base units may refer to other base units, i.e., both independent and dependent definitions of quantity units are possible.

It is interesting that the historical evolution of systems of units has mainly concerned the type of definition adopted, with a trend from non-scaled based systems involving independent (T1) definitions to scaled based systems involving dependent (T3) definitions.[12] In particular, the current SI is a scaled based systems involving all the three types of definitions, while the future revisions of the

SI put forward both a scaled based system involving dependent (**T2**) and (**T3**) definitions and a revolutionary scaled global system involving dependent (**T3**) definitions only.

4.2 A simple system of units

With the aim of exemplifying the various types of definitions, let us introduce a simple system of units with just three base units, for mass M, time T and length L. The definitions are as follows (see BIPM, 2006):

> The kilogram, kg, is the mass of the international prototype of the kilogram:
>
> $$kg = [M] := M(K)$$
>
> The second, s, is the time of 9 192 631 770 periods of the radiation corresponding to the transition between the two hyperfine levels of the ground state of the caesium 133 atom:
>
> $$s = [T] := 9\ 192\ 631\ 770\ \nu(hfs\ Cs)^{-1}$$
>
> The metre, m, is the length of the path travelled by light in vacuum during a time interval of $1/299\ 792\ 458$ of a second:
>
> $$m = [L] := c\ s\ /\ 299\ 792\ 458$$

The reference quantities for the definitions of kg, s and m are then, respectively, the mass of a prototype, the period of a radiation, and both the speed of light in a vacuum and the unit of time.

This system is interesting because:

i   it is an independent subsystem of the current SI;
ii all three types of definitions of units, (**T1**), (**T2**), and (**T3**), are exemplified; and
iii it is a scaled based system including both independent and dependent definitions.

The structure of these definitions is seen in Figure 4.2.

Each arrow represents the relation between a reference quantity and the unit it defines.

Once the reference quantities have been specified, only scale factors remain to be chosen, e.g., 9 192 631 770 in the definition of s.

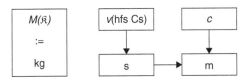

Figure 4.2 The structure of the simple system of units.

Note that this structure is hybrid: some definitions are independent, and some are dependent, giving rise to a hierarchical structure in which the independent definitions are the basis for the dependent ones. Whether a hybrid structure is the best one is a topic that we will discuss in the following.

4.3 Analysis of types of definition

Let us analyse these definitions by considering their structure, the hypotheses on which they are based and the related drawbacks.

The kilogram is defined by reference to the quantity of an object, thus according to a definition of type (**T1**). It is a non-scaled independent definition whose *structure* is:

$$[Q] := Q(s^*)$$

where s^* is an object (the international prototype of the kilogram) and $Q(s^*)$ is the quantity under consideration (its mass). Hence by definition $Q(s^*) = 1[Q]$, so that s^* realizes $[Q]$ with null uncertainty.

This definition is based on the following *hypothesis*: the unit is defined in a given condition c_{DEF} of s^* at a given time t_{DEF}, i.e., kg $:= Q(s^*,c_{DEF},t_{DEF})$: for all times t, it is assumed that $Q(s^*,c_{DEF},t) = Q(s^*,c_{DEF},t_{DEF})$. The quantity Q of s^* in the condition c_{DEF} is then hypothesized to be invariant with respect to variations of time, corresponding to the assumption that s^* is a fundamental object and $Q(s^*)$ a fundamental quantity. This definition has the following *drawbacks*: the reference object is bound to space-time, and the production of a standard is necessarily the reproduction of the primary standard s^*; in addition, the Q-related invariance of s^* has no theoretical justification.

The second is defined by reference to the quantity of a kind of object, thus according to a definition of type (**T2**). It is a scaled independent structural definition whose *structure* is:

$$[Q] := Q(S)/\{q\}$$

where S is a kind of object (the given radiation), $Q(S)$ is its considered quantity (its period), and $\{q\}$ is a numerical factor ($9\,192\,631\,770^{-1}$). Hence if s is an object of the kind S then by definition $Q(s) = \{q\}[Q]$, so that each $s : S$ realizes $\{q\}[Q]$ with null uncertainty.

This definition is based on the following *hypothesis*: the unit is defined in a given condition c_{DEF} of $s : S$ at a given time t_{DEF}, i.e., s $:= Q(s,c_{DEF},t_{DEF})$: for all objects $s' : S$, and all times t, it is assumed that $Q(s',c_{DEF},t) = Q(s,c_{DEF},t_{DEF})$. The quantity Q of $s' : S$ in the condition c_{DEF} is then hypothesized to be invariant with respect to variations of time, corresponding to the assumption that S is a fundamental kind of object and $Q(s)$ a fundamental quantity. The reference is not bound to space-time, and the production of a standard is the realization of an object of the kind S, affected by a realization uncertainty. The Q-related invariance of S is tested empirically. There are no principled *drawbacks* here: if the hypothesis concerning the invariance of $Q(S)$ is justified, the definition provides a stable reference quantity for the unit.

The metre is defined by reference to a fundamental constant quantity, according to a definition of type (**T3**). It is a scaled dependent constant definition, whose *structure* is:

$$[Q] := (k/\{k\})\ \Pi_i[Q_i]^{n_i}$$

where k is a fundamental constant quantity (the speed of light in vacuum), $\{k\}$ is a number that, as a consequence of the definition of $[Q]$, becomes the numerical value of k (299 792 458), and each $[Q_i]$ is a unit raised to the exponent n_i (there is only one unit here, s, raised to the 1st power).

This definition is based on the following *hypothesis*: k is a fundamental constant.

Note that a (**T3**) definition can be interpreted as a (**T2**) definition by introducing an intermediate, independent definition for a unit of a derived quantity. For example, in the case of the metre a unit for speed could be introduced:

$$u = [V] := c\ /\ 299\ 792\ 458$$

so that

$$m := u \cdot s$$

coherently defines the metre in function of the units of speed and time.

The structure of the definition then becomes:

$$[Q] := \Pi_i[Q_i]^{n_i}$$

thus highlighting that in this system the metre is actually a derived unit and the unit of speed is a base unit, even though the metre is presented as a base unit and the unit of speed is not even given a name in the system. There are no principled *drawbacks* here: if the hypothesis concerning the invariance of the constants involved is justified, the definition provides a stable reference quantity for the unit.

5   Analysis of definitions

We can now analyse in a more detailed way the types of definitions exemplified earlier.

5.1   *Definitions based on objects*

The definitions of units as quantities of one or more objects (**T1**) are the most problematic ones. This can be shown by comparing the way of reference to particular objects and to kinds of objects, and then analysing how these ways of reference are involved in the process of unit definition. We can refer to a particular object *by address*, so that the intended quantity is identified by pointing to an object as the support of that quantity. Thus, the unit of length might be defined as the length of a particular object, possibly multiplied by a numerical constant. A suitable definition is then essentially dependent on the hypotheses assumed on the traits and the stability of the traits of the referred object. For example, we might have a model according to which the Earth is (a) perfectly and invariantly spherical, (b) perfectly and invariantly homogeneous and (c) endowed with a perfectly uniform rotatory motion. Such a model would justify referring to the Earth to define the units of length, mass, and time. However, macroscopic objects are usually subject to change, so that they could be characterized by different individual quantities in different times. This problem is clearly discussed by Maxwell:

> The Earth has been measured as a basis for a permanent standard of length, and every property of metals has been investigated to guard against any alteration of the material standards when made. To weigh or measure any thing with modern accuracy, requires a course of experiment and calculation in which almost every branch of physics and mathematics is brought into requisition. Yet, after all, the dimensions of our Earth and its time of rotation, though, relatively to our present means of comparison, very permanent, are not so by any physical necessity. The Earth might contract by cooling, or it might be enlarged by a layer of meteorites falling on it, or its rate of revolution might slowly slacken, and yet it would continue to be as much a planet as before.
>
> (1870, p. 225)

We refer to a kind of object *by definition*, so that the intended quantity is identified by specifying the kind itself, as characterized in terms of a given set of quantities and physical interactions, in such a way that each object of that kind is assumed as the support of that quantity. In this sense a kind of objects can be introduced only if the existence of objects is assumed that are indistinguishable with respect to the quantities that characterize the kind. Again, according to Maxwell:

> A molecule, say of hydrogen, if either its mass or its time of vibration were to be altered in the least, would no longer be a molecule of hydrogen. If, then, we wish to obtain standards of length, time, and mass which shall be absolutely permanent, we must seek them not in the dimensions, or the motion, or the mass of our planet, but in the wave-length, the period of vibration, and the absolute mass of these imperishable and unalterable and perfectly similar molecules.
>
> (Maxwell, 1870, p. 225)

These differences, and the problems related to the first type of definition (**T1**), push towards the idea of focusing on definitions based on references to kinds of objects and fundamental constants.

5.2 Definitions based on kinds of objects or on constants

Definitions based on kinds of objects (**T2**) or on constants (**T3**) are preferable from a theoretical point of view, since they refer to entities that are considered stable according to theories – the best current ones. A further analysis of the structure of these definitions shows that they can be articulated in at least three different versions, which we will call, in accordance with the current literature (see Mills, Mohr, Quinn, Taylor, & Williams, 2006, pp. 232–233):

i   Explicit Unit Definition (EUD);
ii   Explicit Constant Definition (ECD); and
iii   Global Explicit Constant Definition (GCD).

As we will see, the third version is particularly interesting, as it allows us first to provide a definition of the units in a system without referring to any distinction between base vs. derived quantities, and second to disentangle the definition of a unit from the reference to a specific procedure of realization of the unit itself.

5.2.1 Explicit Unit Definitions (EUD)

The schema of an Explicit Unit Definition is:

(EUD) $[Q] := Q(S) = Q(s)$, where $s : S$

being S a kind of entity specified according to a given condition.

Example 1: the metre is the distance travelled by light in vacuum in an interval of time of k seconds.

The instance of the schema is:

$m = [L] := L(s)$ where $s : <x \mid L(x) = c \cdot k \cdot s>$

In this case c is the speed of light in a vacuum, $L(x)$ is the length of x and $<x \mid L(x) = c \cdot k \cdot s>$ is the kind of objects whose length is equal to the distance travelled by light in a vacuum in an interval of time of k seconds. Since s is an entity of that kind, the definition can be rewritten as:

$m = [L] := L(s)$ where $L(s) = c \cdot k \cdot s$
$m = [L] := c \cdot k \cdot s$

so that $k = (1/c) \, m \cdot s^{-1} = (1/\{c\}) \, [c]^{-1} \cdot m \cdot s^{-1} = 1/\{c\}$.

The definition is dependent and involves an implicit reference to a physical constant, the speed of light in vacuum c.

Example 2: The kilogram is the mass of a body whose equivalent energy is equal to that of a number of photons whose frequencies sum to k hertz.

The instance of the schema is:

$$kg = [M] := M(s) \text{ where } s: <x \mid E(x)/h = k \cdot s^{-1}>$$

In this case h is the quantum of action, $M(x)$ is the mass of x, $E(x)$ is the energy of x and $<x \mid E(x)/h = k \cdot s^{-1}>$ is the kind of objects whose equivalent energy is equal to that of a number of photons whose frequencies sum to k hertz. Since s is an entity of that kind, and given the law $E = mc^2$, the definition can be rewritten as:

$$kg = [M] := M(s) \text{ where } s: <x \mid M(x) \cdot c^2/h = k \cdot s^{-1}>$$
$$kg = [M] := M(s) \text{ where } M(s) \cdot c^2/h = k \cdot s^{-1}$$
$$kg = [M] := k \cdot h/c^2 \cdot s^{-1}$$

so that $k = c^2/h \cdot kg \cdot s = \{c\}^2/\{h\} \cdot [c]^2 \cdot [h]^{-1} \cdot kg \cdot s = \{c\}^2/\{h\}$.

This definition is implicitly dependent on both the definition of the unit of time and the definition of the unit of length. In addition, it depends on the assumption of the validity of physical laws concerning the mass-energy equivalence, $E = mc^2$, and the quantization of energy, $E = hv$. Hence, the definition involves a reference to two physical constants, the speed of light in vacuum c and the quantum of action h.

The analysis of these examples shows that:

i   a unit is defined by virtue of an explicit reference either to an instance of a quantity of the same kind or to an instance of a quantity of a different kind which is connected to instances of quantities of the same kind of the unit by means of a physical law;

ii in the definition the reference to other units is *explicit*, so that the hierarchy of the definitions proposed in a system appears from the structure of the definitions themselves;

iii in the definition the reference to physical constants is left *implicit*, so that the specific relational structure of the definition, depending on the dimensionality of the constants, is not apparent;

iv in the definition the reference to physical laws is *essential* and is left *implicit*, so that the specific dependence of the definition from the laws is not apparent; and

v   the articulation of the definition does not completely reflect the structure of the definition.

5.2.2   *Explicit Constant Definitions (ECD)*

The schema of an Explicit Constant Definition is:

$$(ECD) \; [Q] \text{ is such that } \kappa = k \prod_i [Q_i]^{n_i}$$

where $\dim \prod_i [Q_i]^{n_i} = \dim Q$, κ is a fundamental constant quantity and k is a numerical constant.

Example 1: the metre is the length such that the numerical value of the speed of light in vacuum c is k metres per second.

The instance of the schema is:

$$m = [L] \text{ is such that } c = k \cdot m \cdot s^{-1}$$

As a consequence, we obtain that $\{c\} = k$, since $c = \{c\} \cdot m \cdot s^{-1}$.

The definition depends on the definition of the unit of time and makes explicit reference to the physical constant c.

Example 2: the kilogram is the mass such that the numerical value of the quantum of action h is k kilograms metre^2 per second.

The instance of the schema is:

$$kg = [M] \text{ is such that } h = k \cdot kg \cdot m^2 \cdot s^{-1}$$

As a consequence, we obtain that $\{h\} = k$, since $h = \{h\} kg \cdot m^2 \cdot s^{-1}$.

The definition depends on the definition of the units of time and of length and makes explicit reference to physical constant h.

The analysis of these examples shows that:

i   a unit is defined by virtue of an explicit reference to a constant;

ii in the definition the reference to other units is left *implicit*, so that the hierarchy of the definitions proposed in a system does not appear from the structure of the definitions themselves;

iii in the definition the reference to the physical constants is *explicit*, so that the specific relational structure of the definition, depending on the dimensions of the constants, is apparent;

iv in the definition the reference to the physical laws is *not essential*: one can use whatever laws in order to check whether the value of the constant is the prescribed value; and

v   the articulation of the definition completely reflects the structure of the definition. Indeed, the general structure of any definition of this kind is:

$$u, \text{ unit of } Q, \text{ is such that } \{\kappa\} = k.$$

Accordingly, Mills et al. conclude:

> Such a definition has the advantages that it is simple, concise and makes clear the fundamental constant to which the unit is linked and the exact value of that constant. If this general form were chosen, it would be appropriate to choose definitions of the same form for all seven base units.
>
> (2006, p. 234)

Finally, it is not difficult to see that EUD and ECD are equivalent, once we are provided with a specific set of laws for determining the value of the physical

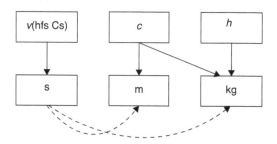

Figure 4.3 Dependence relations in the definitions of units.

constant. For example, in the case of the definition of the unit of length, we get $m = c \cdot k \cdot s \Leftrightarrow c = (1/k) \cdot m \cdot s^{-1}$, since $k = 1/\{c\}$. However, moving from EUD to ECD highlights the reference to a set of constants, i.e., the set of quantities that, by definition, are settled as *constant with null uncertainty*. Hence, in ECD, the distinction between quantities that are fixed as constant on the basis of the assumption of a certain theoretical framework, constituted by a set of physical laws, and quantities that are not so fixed emerges in the best possible way. Actually, in ECD, to fix a unit coincides with fixing the numerical value of a constant in a given setting.

Both in EUD and in ECD, the dependence relations in the definitions are shown in Figure 4.3.

As we can see, the definitions are highly interdependent, and the same constant can be used in the definition of different units.

5.2.3   Global Constant Definitions (GCD)

In ECD constants are essential elements in local definitions of units, as every definition fixes a unique unit, in a setting where other units and constants possibly occur. However, a set of units can be completely specified once the numerical value of a corresponding number of physical constants is fixed, provided that the dimensions of these fundamental constant constitute a base of the space of the dimensions (see Mills et al., 2006, pp. 234–235; Mohr, 2008; Cabiati & Bich, 2009). Hence, the possibility is open to define a system of units in one shot, by fixing once and for all the numerical values of a set of fundamental constants.

The schema of a Global Constant Definition is then:

(GCD) $[Q_i]$, $i = 1, \ldots, N$, is a system of units such that $\{\kappa_i\} = k_i$

where each κ_i is a fundamental constant quantity, each k_i is a numerical constant and N is the number of elements of a base of the space of the dimensions in the given system of quantities.

In this case, the units of all quantities are simultaneously defined by virtue of an explicit reference to a set of fundamental constants, so that the distinction between base and derived units is discarded:

> A major advantage of the proposed new approach is that it does away entirely with the need to specify base units and derived units and hence the confusion that this requirement has long been recognized to engender, not the least of which is the arbitrariness of the distinction between base units and derived units. This need is eliminated by no longer having a unique, one-to-one correspondence between a particular unit and a particular fundamental constant.
>
> (Mills et al., 2006, p. 235)

In addition, it can be noted that GCD is equivalent to any set of local definitions (ECD), provided that the same set of physical constants is assumed as the basis of definition, and that it allows for procedures of realization of units based on any physical law where the unit whose instance is to be realized is suitably connected with the chosen set of fundamental constants. Further advantages are well highlighted by Cabiati and Bich:

> Fundamental constants, being universal and invariant quantities, are ideal references for measurement units. In this role they are preferable to other physical quantities, such as a parameter of a particular natural substance in a specified state or even of a specific atom, which are equally invariant, but have no general role in scientific theories. Assigning exact values to a suitable subset of fundamental constants can significantly reduce the uncertainties of several other related constants, thus resulting in a considerable impact on the scientific use of the SI. Some advantage can also come from the coherence of the system, which is a property not always guaranteed by the unit definition when it is based on a reference quantity with no theoretical relations with other quantities.
>
> (2009, p. 457)

In synthesis, GCD seems to be the best candidate for providing a consistent definition of a system of units. Still, in the case of GCD all units are defined in a dependent way, and therefore the modularity of the system of definitions is lost. This calls for a further, final discussion.

6   Conclusions: assessment of the types of definition

Let us sum up the desiderata that an ideal system of units should satisfy. The definition of the units in the system should guarantee:

i   the *stability* of the realizations, i.e., their invariance with respect to variations of space-time regions, with minimal uncertainty;

ii   the *intersubjectivity* of the realizations, i.e., their invariance with respect to the agents realizing them, with minimal uncertainty; and

iii the *continuity* with the units that are traditionally used.

In order to check how the different strategies satisfy these desiderata, two further decision criteria can be taken into account:

i   *temporally local vs. temporally global revision strategy*, whether any possibly required change should be consistently performed one definition at a time or all definitions at once; and

ii   *non-modular vs. modular revision strategy*, whether any possibly required change should be consistently performed so that a change in one definition produces effects on the whole set of definitions or on the directly involved unit only.

6.1   Temporally local vs. temporally global revisions

The revision of a system of units might follow a simple step-by-step process: a set of definitions is introduced based on a set of reference quantities; whenever variations in such reference quantities are discovered the definitions are revised based on a different set of reference quantities, and so on. In this process the original reference quantities are initially considered stable, either because no procedures are available to test possible variations or because, even in the presence of testing procedures, the variations fall in an acceptable uncertainty interval. The discovery of variations in the reference quantities is then possible as a consequence of advances in measurement. This kind of process, triggered by the discovery of variations in the reference quantities and of more stable quantities, is then temporally local: whenever the discovery occurs, a new definition is introduced. In this case, there is no principled consideration requiring to globally revise the system of units. The opposite case is a definition of the system of units in which a set of reference quantities is selected once and for all. The problem here is evidently to correctly identify the set of stable quantities, since such an identification crucially depends on the current scientific framework. Still, *ceteris paribus*, a temporally global revision process seems to be better than a temporally local revision process, since the consistency of the resulting system is an expected constraint on a global process.

Among the types of definitions discussed earlier, GCD is the one that better fits this global revision strategy. In fact, in a GCD system all units are simultaneously defined and are defined once and for all. Furthermore, in a GCD system both criteria (**C1–C3**) and (**C4–C6**) (see Section 3.1) are best satisfied, since fundamental constants are stable by definition in a given theoretical framework, and definitions are independent of specific procedures of realization, so that the best available procedures can be adopted on a case-by-case basis.

6.2 A possible development

In principle, the definition of a system of units could be changed in response to two kinds of problems. On the one hand, as a consequence of a change of the current physical theories, we might discover that some of the quantities that are presently considered constant are not actually constant. On the other hand, as a consequence of a change of the current knowledge, we might discover that to some of the quantities to which given values are presently assigned new values have to be assigned. The first kind of change is radical, since the discover of new relations between physical constants would imply a revision of the entire system. The second kind of change is less extreme, and it is interesting to consider how a similar change could be handled depending on the way in which the definition of the system is articulated. In particular, while in both cases the system of units should change to accommodate the new data, we deem that the only way for a GCD system, as presented, to be adapted is to change the quantities of the units, so that a change due to a scientific discovery would lead to a discontinuity in the set of units. Let us develop this point, which will lead us to suggest a possible improvement of the GCD strategy.

6.2.1 A drawback

In a GCD system units are defined by taking a set $\{C_i\}$ of physical constant quantities as reference quantities and fixing a given numerical factor k_i for each C_i. Every unit $[Q]$ is then a product of powers of the constant quantities C_i:

$$[Q] = \prod_i (C_i / k_i)^{n_i}$$

Hence, a system of units S is defined as the system of units where the constant quantities C_i take values k_i.

Since, according to the best available physical theories, fundamental constants are universal and invariant quantities, they represent an ideal reference for defining a system of units. Still, if so defined, a system of units is crucially dependent on the set $\{k_i\}$, a condition that might become a drawback.

In order to see the potential problem deriving from plugging $\{k_i\}$ into the very definition of the system, let us note, first of all, that the constant quantities C_i do not only operate as connectors among quantities in fundamental physical laws but also are quantities of kinds of objects. For example, the constant quantity c not only connects energy and mass in the equation $E = mc^2$, but is also the speed of light in a vacuum. This second function makes the constant quantities C_i in principle directly measurable, through an empirical interaction with their kind of objects, for example light in a vacuum in the case of c. Furthermore, it is by means of measurement that the numerical value k_i has been identified for C_i, and then maintained in the definition of the system of units S according to the principle of continuity. This generates the principled possibility that a systematic error was made in the measurement.

Of course, before a system of units is defined a quantity cannot be measured, but even without a unit quantities of the same kind can be compared. Let us then consider the following, admittedly imaginary, scenario. Suppose that (a) our Solar System is in a somewhat special region of the universe, characterized by the fact that one of the assumed fundamental constants, say C_1, has the numerical value k_1 (in the appropriate unit), as measured before the introduction of the system of units S, and that (b) the value of C_1 would be different, say k_1^*, if measured in any other region of the universe. Before the introduction of S this discovery would just lead us to conclude that the fundamental constant C_1 has the numerical value k_1^*, in the appropriate unit, thus correcting the previous value. If instead the discovery is done after the introduction of S, and S is indeed defined in terms of $\{k_i\}$, we face a dilemma, since we have to opt for:

i   preserving the system, and so admitting that S, after all, is not based on a set of fundamental constants; or

ii   changing the whole definition, and moving to a new system S^*, where the value associated to C_1 is k_1^*.

This is so because we lack the possibility of intervening on the system by changing only one parameter.

6.2.2   A solution

Let us see how a modification of the GCD strategy can allow us both to provide a solution to the foregoing problem and to better understand the structure of the GCD definition itself.

In the development of a system of units, we note the presence of two independent trends. First, from constant quantities of objects or kinds of objects to constant quantities related to physical laws. Second, from a setting where the distinction of base and derived quantity is essential to a setting where this distinction, while still useful to identify dimensions of quantities, is less significant.

In presenting the definition of a system of units the distinction between base and derived quantities is indeed not completely disregarded, since the very specification of the system includes a reference to a certain set of units. Hence, in the example introduced in Section 4.2, the definition is to be given as follows:

> The International System of Units, the SI, is the system of units scaled so that (1) the ground state hyperfine splitting transition frequency of the caesium 133 atom $\Delta\nu(^{133}\text{Cs})_\text{hfs}$ is 9192631770 hertz; (2) the speed of light in vacuum c is 299792458 metres per second; (3) the Planck constant h is $6.6260693 \times 10^{-34}$ joule second.
>
> (Mills et al., 2006, pp. 235–236)

For the definition to be effective, the terms "hertz", "metre", "second" and "joule" have to be understood, implying the knowledge that they refer to a unit of frequency, length, duration and energy respectively. Hence, the previous definition should be articulated as follows:[13]

> The International System of Units, the SI, is the system of units scaled so that (1) the ground state hyperfine splitting transition frequency of the caesium 133 atom $\Delta\nu(^{133}Cs)_{hfs}$ is 9192631770 units of frequency; (2) the speed of light in vacuum c is 299792458 units of length per unit of time; (3) the Planck constant h is $6.6260693 \times 10^{-34}$ units of energy times unit of time.

This implies that, in providing the definition, a certain selection of units has to be made, so that a distinction between base units and derived units is recoverable from the definition itself. It is peculiar that the definition involves units of quantities of a basis that does not coincide with the natural basis induced by the defining constants. Indeed, in view of the fact that the system of units is to be a coherent system, we could simply say:

> The International System of Units, the SI, is the coherent system of units scaled so that (1) the ground state hyperfine splitting transition frequency of the caesium 133 atom $\Delta\nu(^{133}Cs)_{hfs}$ is 9192631770 units of frequency; (2) the speed of light in vacuum c is 299792458 units of velocity; (3) the Planck constant h is $6.6260693 \times 10^{-34}$ units of action.

In fact, we know, independently of the definition of the system of units, how to represent any set of units in any chosen basis. Finally, this definition can be suitably structured so that the distinction between *natural units* and *conventional units* becomes explicit:

> The Natural System of Units, the SN, is the coherent system of units where (1) the ground state hyperfine splitting transition frequency of the caesium 133 atom $\Delta\nu(^{133}Cs)_{hfs}$ is the unit of frequency; (2) the speed of light in vacuum c is the unit of velocity; (3) the Planck constant h is the unit of action.
>   The International System of Units, the SI, is the coherent system of units, based on time, length, and mass, scaled so that (1) the unit of time is 1/9192631770 of the natural unit of frequency; (2) the unit of length is 1/299792458 of the natural unit of velocity times the unit of time; (3) the unit of mass is the mass of a body whose equivalent energy is equal to that of a number of photons whose frequencies sum to $(299792458^2 / 66260693) \cdot 10^{41}$ natural unit of frequency.

Consequently, the definition of a system of units might be given along the following lines:

i S is the system of units where C_i takes the *natural* value 1
ii   which is scaled, for continuity, by a factor $v(C_i)$ so that
iii S is the system of units where C_i takes the *conventional* value $v(C_i)$
iv   which, to the best of our current knowledge, is fixed to k_i.

In addition, since the units can be defined in any dimensional basis, we are free to select as basic kinds of quantity the ones that fits at best with the continuity criterion.

In this way, the fundamental choices on which the construction of a system of units rests are all explicit. These choices are: (a) the selection of the constants; (b) the selection of parameters for scaling the system of units; and (c) the selection of a base for the system of units itself.

Step 1 – Selection of the constants: the number of the constants has to coincide with the number of independent kinds of quantity in the system of quantities, since they have to provide a basis for the corresponding space. The choice of the constants is indeed a choice of kinds of quantity, since each constant is a quantity of a kind. In principle this is also a possible choice of units, since each constant could be assigned the value 1.

Step 2 – Selection of the scale parameters: the parameters are chosen in such a way that continuity is preserved. The choice of the parameters rescales the basis for the corresponding vector space and is driven by the constraint of preserving the traditional units, even though with a radically new definition.

Step 3 – Selection of the base kinds of quantity: the number of the base kinds of quantity has to coincide with the number of independent kinds of quantity in the system of quantities, since they have to provide a basis for the corresponding vector space. The choice of the base kinds of quantity defines a new basis for the vector space and is made so as to preserve continuity with the current set of base units.

Consequently, we get both that a change concerning the size of a fundamental constant turns out to be possible and that such a change would only induce a change in the known value of a conventional scale factor. This shows that such a system is *structurally rigid*, as obtained in Step 1, due to the expected stability of fundamental scientific knowledge, so that only *theory changes* would require interventions on its structure, but at the same time, *partly modular*, due to the parameterization of the structure. This strategy would effectively cope with the discontinuities highlighted earlier, while ensuring the basic condition of social continuity of units.

The proposed structure of definition provides us with two more benefits.

First, *it emphasizes the distinction between what is claimed to be based on theory and what is instead conventional in the definition*. In the current SI, natural units are acknowledged and listed (BIPM, 2006, Table 7). According to the *SI Brochure*, natural units are significant, because in many fields "it is most convenient to express the results of experimental observations or of theoretical calculations in terms of fundamental constants of nature" [p. 125], but their use has not to be encouraged, because "the quantity systems on which these units are based differ so fundamentally from that on which the SI is based" [p. 125]. Hence, natural units are "units whose values in SI units have to be determined experimentally, and thus have an associated uncertainty" [p. 125]. Now, the adoption of a GCD strategy is also motivated by the effort to eliminate the uncertainty of a number of fundamental constants, and fundamental constants play an essential role in such a strategy. Therefore, a definition in which this role is recognized, and the use of natural units is legitimized, seems to be appropriate.

Second, *the proposed structure appears more easily presentable to and understandable by a wide readership, even without a specific background in physics.* In fact, it motivates (a) the reference to fundamental constants, which are explicitly presented as a natural choice, being natural units, (b) the choice of scale factors, which are to be introduced because of continuity, and (c) the choice of precisely those scale factors, which are the ones determined according to our best knowledge.

Given the critical role that the SI plays in our society, this seems to be an important condition to fulfil.

Notes

1 While the *International Vocabulary of Metrology* (JCGM, 2012) uses the term "measurement unit", "quantity unit" is at the same time more general (units can be used also in evaluations that are not measurements) and more flexible ("quantity unit" can be specialized as, e.g., "length unit", "mass unit", where length, mass, are quantities, not measurements). In the present context "unit" is sometimes used instead of "quantity unit" for the sake of brevity.

2 See Mills, Mohr, Quinn, Taylor, and Williams (2011) and Newell (2014) for an introduction to the general problems involved in defining and improving a system of units and for a presentation of the current state of the debate. See Milton, Davis, and Fletcher (2014) for a review of the advances made in the recent years.

3 A general reference on basic concepts of metrology is the already mentioned *International Vocabulary of Metrology* (JCGM, 2012). A model of measurement based on the conceptual framework presented here is proposed in Frigerio, Giordani, and Mari (2010).

4 See Mari (2009) and Mari and Giordani (2012) for further developments.

5 A system of quantities actually relates kinds of quantity, so that a better term would be "system of kinds of quantity". The same can be said about dimensions of quantities, base quantities, etc., that are indeed dimensions of kinds of quantity, base kinds of quantity, etc. We will maintain the usual, shorter terms here.

6 Once $\mathbf{B}$ is selected, base kinds of quantity are identified with kinds of quantity in $\mathbf{B}$, and derived kinds of quantity with kinds of quantity in $\mathbf{Q} - \mathbf{B}$.

7 Here δ_{ij} is the Kronecker delta, $\delta_{ij} = 1$ if $i = j$ and $\delta_{ij} = 0$ otherwise. If Q_i is a base quantity, then all exponents in its representation in $\mathbf{B}$ are zero except for $i = j$.

8 Strictly speaking, the kind of quantity T is duration, not time (analogously L is indeed length, not space). We maintain the term "time" only by habit.

9 We follow the usual notation for a quantity value in terms of a number and the unit. The slight modification in the denotation of the individual quantity and the numerical value, for which a lower case letter is used, is due to the fact that we use upper case letter to denote kinds. The unit is still denoted by $[Q]$, since it is properly referred to a kind of quantity. The notation $q = \{q\}[Q]$ can be interpreted as follows. The selection of $[Q]$ induces a function $\{\}_{[Q]} : Ext(Q) \to \mathbf{R}^+$ such that $\{q\}_{[Q]} = q / [Q]$. Hence, to obtain a complete representation of q it is sufficient to define a function $[] : \{Q\} \to Ext(Q)$, which selects an individual quantity $[Q] \in Ext(Q)$.

10 We do not discuss here the issue of the possibility of ordinal or nominal measurement. On this matter see Mari and Giordani (2012).

11 See Mills et al. (2011), pp. 227–228.

12 For a synthetic overview of the evolution of systems of units, see Cabiati (2015), § 2.3 and Leschiutta (2007).
13 This definition still includes a manifest problem of the current hypothesis about the "New SI": one reference quantity (the transition frequency of the caesium) is clearly not a fundamental constant but the quantity of a kind of object. On the other hand, the discussion of this problem is not relevant to the scope of this chapter.

References

BIPM. (2006). *SI Brochure: The International System of Units (SI)* (8th ed.), updated in 2014. Paris: Bureau International des Poids et Mesures. Retrieved from www.bipm.org/en/publications/si-brochure.

BIPM. (2014). *On the future revision of the SI*. Paris: Bureau International des Poids et Mesures. Retrieved from www.bipm.org/en/measurement-units/new-si.

Cabiati, F. (2015). The system of units and the measurement standards. In A. Ferrero, D. Petri, P. Carbone, & M. Catelani (Eds.), *Measurements: Fundamentals and applications* (pp. 47–84). Hoboken, NJ: Wiley-IEEE Press.

Cabiati, F., & Bich, W. (2009). Thoughts on a changing SI. *Metrologia, 46*, 457–466.

Frigerio, A., Giordani, A., & Mari, L. (2010). Outline of a general model of measurement. *Synthese, 175*, 123–149.

JCGM. (2012). *International vocabulary of metrology: Basic and general concepts and associated terms* (VIM) (3rd ed.). Joint Committee for Guides in Metrology. Retrieved from www.bipm.org/en/publications/guides/vim.html.

Leschiutta, S. (2007). History of standard definitions: An outline. In T. Haensch, S. Leschiutta, & A. J. Wallard (Eds.), *Metrology and fundamental constants* (pp. 45–58). Amsterdam: IOS Press.

Mari, L. (2009). On (kinds of) quantities. *Metrologia, 46*, L11–L15.

Mari, L., & Giordani, A. (2012). Quantity and quantity value. *Metrologia, 49*, 756–764.

Maxwell, J. C. (1870). Address to the Mathematical and Physical Sections of the British Association. In *The scientific papers of James Clerk Maxwell* (Vol. 2, pp. 215–229). Cambridge: Cambridge University Press.

Mills, I. M., Mohr, P. J., Quinn, T. J., Taylor, B. N., & Williams, E. R. (2006). Redefinition of the kilogram, ampere, kelvin and mole: A proposed approach to implementing CIPM recommendation (CI-2005). *Metrologia, 43*, 227–246.

Mills, I. M., Mohr, P. J., Quinn, T. J., Taylor, B. N., & Williams, E. R. (2011). Adapting the International System of Units to the twenty-first century. *Philosophical Transaction of the Royal Society* A, *369*, 3907–3924.

Milton, M. J. T., Davis, R., & Fletcher, N. (2014). Towards a new SI: A review of progress made since 2011. *Metrologia, 51*, R21–R30.

Milton, M. J. T., Williams, J. M., & Bennett, S. J. (2007). Modernizing the SI: Towards an improved, accessible and enduring system. *Metrologia, 44*, 356–364.

Mohr, P. J. (2008). Defining units in the quantum based SI. *Metrologia, 45*, 129–133.

Newell, D. B. (2014). A more fundamental International System of Units. *Physics Today, 67*, 35–41.

5 Relations between units and relations between quantities

Susan G. Sterrett

1 Introduction

1.1 Background

As this chapter goes to press, it is now assured that we will soon have in place an international system of units in which *all* the units of the SI (Système International d'Unités) are defined in terms of "well-recognized fundamental constants of nature" (BIPM, 2014; BIPM, 2011b; Mills, Mohr, Quinn, Taylor, & Williams, 2011; Quinn 2019; Bordé 2019). The specific change most discussed so far has been that the definition of the unit for the quantity mass, i.e., the kilogram, is to change, and that it is to change in such a way that it will no longer be defined in terms of a physical artefact (Riordan 2011; Quinn 2019).

Changing the definition of the kilogram in this way also changes the network of interrelations amongst the definitions of the base SI units. After the proposed change, the definition of the kilogram, which is currently not dependent on the definition of any of the other base SI units, becomes dependent on, or interrelated with, the definitions for some of the other SI units. The kilogram is not the only unit whose definition will be changed by the proposed change to the SI, though: four of the base units of the SI will be redefined in terms of "constants of nature". These changes in *how* the base units are defined have been discussed, too; many such discussions focus on *how units in the SI are related to the "constants of nature"* in terms of which they are defined. I will be more interested here in how the proposed changes to the definitions of SI units impact *the relations between units of the SI.*

1.2 Definitions of units in the SI

There are other differences[1] between the proposed SI and the current SI that are even more fundamental, in that they have to do with the *means by which* the units are defined, i.e., the way in which the definitions are formulated (as is also mentioned in the chapters by Quinn and Bordé introducing this book). For completeness, I provide in this chapter three tables showing the definitions of units in the current SI (Table 5.1) and in the proposed "New SI" (Tables 5.2 and 5.3), for reference in the discussion later on.

Table 5.1 Base quantities and base units used in the (current) SI

Base quantity	Base unit	Symbol for unit	Definition of unit
length	metre	m	The metre is the length of the path travelled by light in vacuum during a time interval of 1/299 792 458 of a second.
mass	kilogram	kg	The kilogram is the unit of mass; it is equal to the mass of the international prototype of the kilogram.
time, duration	second	s	The second is the duration of 9 192 631 770 periods of the radiation corresponding to the transition between the two hyperfine levels of the ground state of the caesium 133 atom.
electric current	ampere	A	The ampere is that constant current which, if maintained in two straight parallel conductors of infinite length, of negligible circular cross-section and placed 1 m apart in vacuum, would produce between these conductors a force equal to 2×10^{-7} newton per metre of length.
thermo-dynamic temperature	kelvin	K	The kelvin, unit of thermodynamic temperature, is the fraction 1/273.16 of the thermodynamic temperature of the triple point of water.
amount of substance	mole	mol	1. The mole is the amount of substance of a system that contains as many elementary entities as there are atoms in 0.012 kilograms of carbon 12.

2. When the mole is used, the elementary entities must be specified, and may be atoms, molecules, ions, electrons, other particles or specified groups of such particles. |
| luminous intensity | candela | cd | The candela is the luminous intensity, in a given direction, of a source that emits monochromatic radiation of frequency 540×10^{12} hertz and that has a radiant intensity in that direction of 1/683 watt per steradian. |

Source: "A concise summary of the International System of Units, the SI" published by the BIPM, dated March 2006 www.bipm.org

Table 5.2 Definition of "New SI" in terms of seven "defining constants"

The international System of Units, the SI, is the system of units in which

– the unperturbed ground state hyperfine transition frequency of the caesium 133 atom Δv_{Cs} is 9 192 631 770 Hz,

– the speed of light in vacuum c is 299 792 458 m/s,

– the Planck constant h is 6.626 070 15 × 10^{-34} J s,

– the elementary charge e is 1.602 176 634 × 10^{-19} C,

– the Boltzmann constant k is 1.380 649 × 10^{-23} J/K,

– the Avogadro constant N_A is 6.022 140 76 × 10^{23} mol^{-1},

– the luminous efficacy of monochromatic radiation of frequency 540 × 10^{12} Hz, K_{cd} is 683 lm/W.

where the hertz, joule, coulomb, lumen, and watt, with unit symbols Hz, J, C, lm, and W, respectively, are related to the units second, metre, kilogram, ampere, kelvin, mole, and candela, with unit symbols s, m, kg, A, K, mol, and cd, respectively, according to Hz = s^{-1} , J = m^2 kg s^{-2}, C = A s, lm = cd m^2 m^{-2} = cd sr, and W = m^{2}kg s^{-3}.

Source: 26th CGPM Resolution

The change in how the units are defined in the New SI touches on the very notion of the role of the base units in the SI. For, after the proposed change, there will be *two alternative formulations* of the definition of SI units, as follows:

Definition of the New SI without base units

In the first formulation of the definition of the proposed system of SI units (Table 5.3), in an unprecedented move, *no distinction is drawn between base units and derived units among all the SI units mentioned in the formulation*: the "system of SI units" is defined in terms of a collection of SI units that includes not only the seven base units of the current SI (length, mass, time, electric current, thermodynamic temperature, amount of substance and luminous intensity), but also the hertz, joule, coulomb, lumen and watt. The system of units is defined as a whole, in terms of seven constants of nature to be referred to as "the defining constants". The BIPM document describing the proposed change to the SI system is quite clear about there being no distinction between

Table 5.3 Base quantities and base units – proposed revision to the SI Definition of "New SI" in "explicit-constant" formulation

Base quantity	Base unit	Symbol for unit	Definition of unit
time, duration	second	s	The second, symbol s, is the SI unit of time. It is defined by taking the fixed numerical value of the caesium frequency $\Delta\nu_{Cs}$ the unperturbed ground-state hyperfine transition frequency of the caesium 133 atom, to be 9 192 631 770 when expressed in the unit Hz, which is equal to s^{-1}.
length	metre	m	The metre, symbol m, is the SI unit of length. It is defined by taking the fixed numerical value of the speed of light in vacuum c to be 299 792 458 when expressed in the unit m s^{-1}, where the second is defined in terms of the caesium frequency $\Delta\nu_{Cs}$.
mass	kilogram	kg	The kilogram, symbol kg, is the SI unit of mass. It is defined by taking the fixed numerical value of the Planck constant h to be 6.626 070 15 Å $\times 10^{-34}$ when expressed in the unit J s, which is equal to kg m^2 s^{-1}, where the metre and the second are defined in terms of c and $\Delta\nu_{Cs}$.
electric current	ampere	A	The ampere, symbol A, is the SI unit of electric current. It is defined by taking the fixed numerical value of the elementary charge e to be 1.602 176 634 Å $\times 10^{-19}$ when expressed in the unit C, which is equal to A s, where the second is defined in terms of $\Delta\nu_{Cs}$.
thermo-dynamic temperature	kelvin	K	The kelvin, symbol K, is the SI unit of thermodynamic temperature. It is defined by taking the fixed numerical value of the Boltzmann constant k to be 1.380 649 $\times 10^{-23}$ when expressed in the unit J K^{-1}, which is equal to kg m^2 s^{-2} K^{-1}, where the kilogram, metre and second are defined in terms of h, c and $\Delta\nu_{Cs}$.
amount of substance	mole	mol	The mole, symbol mol, is the SI unit of amount of substance. One mole contains exactly 6.022 140 76 Å $\times 10^{23}$ elementary entities. This number is the fixed numerical value of the Avogadro constant, N_A, when expressed in the unit mol^{-1} and is called the Avogadro number.
			The amount of substance, symbol n, of a system is a measure of the number of specified elementary entities. An elementary entity may be an atom, a molecule, an ion, an electron, any other particle or specified group of particles.
luminous intensity	candela	cd	The candela, symbol cd, is the SI unit of luminous intensity in a given direction. It is defined by taking the fixed numerical value of the luminous efficacy of monochromatic radiation of frequency 540 Å $\times 10^{12}$ Hz, *Kcd*, to be 683 when expressed in the unit lmW^{-1}, which is equal to cd srW^{-1}, or cd sr kg^{-1} m^{-2} s^3, where the kilogram, metre and second are defined in terms of h, c and $\Delta\nu_{Cs}$.

Derived units are defined as products of powers of the base units. When the numerical factor of this product is one, the derived units are called *coherent derived units*. The base and coherent derived units of the SI form a coherent set, designated the *set of coherent SI units*.

Source: 2019 version of Draft 9th edition SI Brochure

base and derived units in this formulation; as pointed out in the Draft 9th *SI Brochure* (BIPM, 2013):

> The use of seven defining constants is the simplest and most fundamental way to define the SI. . . . In this way no distinction is made between base units and derived units; all units [mentioned in the definition] are simply described as SI units. This also effectively decouples the definition and practical realization of the units.

To explain what "practical realization of the units" means, I quote from the relevant BIPM document prepared to orient readers to the fundamental change associated with the realization of units in the "New SI":

> In general, the term "to realize a unit" is interpreted to mean the establishment of the value and associated uncertainty of a quantity of the same kind as the unit that is consistent with the definition of the unit. It is important to recognize that any method consistent with the laws of physics and the SI base unit definitions can be used to realize any SI unit, base or derived.
>
> (BIPM, CCE-09-05)

Although the BIPM indicates it will provide specific methods for realizing the units of the New SI, the role of the methods for realizing units provided by the BIPM in the New SI is different than in previous SI systems: "The list of methods given is not meant to be an exhaustive list of all possibilities, but rather a list of those methods that are easiest to implement and/or that provide the smallest uncertainties" (BIPM, CCE-09-05). The result of this "decoupling" of the definition of the units from the practical realization of the units is thus that "while the definitions may remain unchanged over a long period of time, the practical realizations can be established by many different experiments, including totally new experiments not yet devised" (BIPM, 2013, pp. 9–10).

Thus, the proposed system of international units ("New SI") can be defined (a) without drawing a distinction between base units and derived units, and (b) without restricting (or even specifying) the means by which the value of the quantity associated with the unit is to be established. *Both of these are striking changes;* many in philosophy of science will find it surprising that it is possible to specify a system of units without distinguishing base units from derived ones, and without specifying how the unit is to be established. (Or, at least, they will find it thought-provoking to consider how this might be done.) In order to provide a definition of the New SI in a more familiar formulation, an alternative definition has also been provided. The alternative (second) formulation does identify seven SI units as base units.

Definition of the New SI with base units

In the second formulation of the definition of the "New SI":

> [The SI is] defined by statements that explicitly define seven individual base units: the second, metre, kilogram, ampere, kelvin, mole, and candela. These correspond to the seven base quantities time, length, mass, electric current, thermodynamic temperature, amount of substance, and luminous intensity. All other units are then obtained as products of powers of the seven base units, which involve no numerical factors; these are called coherent derived units.
>
> (BIPM, undated)

The definitions of the seven base units in this formulation of the definition of the proposed revision to the SI, or "New SI", are shown in Table 5.2. On this second, alternative, definition of the New SI, each of these seven definitions of an individual base unit is called an "explicit-constant formulation". Unlike in the first formulation of the definition of the "New SI" (shown in Table 5.3), each of the "explicit constant" definitions of a base SI unit identifies a quantity (e.g., time, length, mass, etc.) with which the unit is uniquely associated (Table 5.2) (BIPM, 2011b). For example, the metre is the unit of *length*; the second is the unit of *time*; the ampere is the unit of *electric current*, and so on for the seven base units and the associated seven quantities. Each of the units is then defined in terms of setting its magnitude "by fixing the numerical value of" a certain constant of nature.

While it is not unprecedented to define a unit by fixing the value of a constant of nature for a particular unit (e.g., metre), it is unprecedented to do so for the entire system of units. Those unfamiliar with the background leading up to the decision to take this approach who wish further explanation may find it in chapters 1 and 3 of this book devoted to presenting the New SI. To further aid the reader here and for reference at later points in the paper, I provide, in Table 5.4, a selection from the BIPM's answers to FAQs (Frequently Asked Questions) that I consider especially relevant to the logic of defining units via setting the values of constants of nature.

Even though each of the seven unit definitions in the second, "explicit-constant" formulation of the New SI contains just one of the seven "defining constants", *it is not the case that each base unit of the SI is considered uniquely associated with one of the "defining constants"*. This is clear to see, by inspecting the definitions in Table 5.2: it can easily be seen that most of the definitions of a base unit involve other units, so that often several SI units are jointly involved in using a defining constant to define a unit. Thus, on either definition of the proposed SI system of units, or New SI, the relation between the individual units being defined and the seven "defining constants" is a matter of a *collection* of SI units being *jointly defined* by a *collection of seven constants* considered invariants of nature.

Table 5.4 Selected excerpts from BIPM's "FAQs about the New SI"

FAQ	Question	Answer
8	How can you fix the value of a fundamental constant like h to define the kilogram, and e to define the ampere, and so on? How do you know what value to fix them to? What if it emerges that you have chosen the wrong value?	• We do not fix – or change – the <u>*value*</u> of any constant that we use to define a unit. The values of the fundamental constants are constants of nature and we only fix the <u>*numerical value*</u> of each constant when expressed <u>*in the New SI*</u> <u>*unit*</u>. By fixing its numerical value we define the magnitude of the unit in which we measure that constant. • Example: If c is the *value* of the speed of light, $\{c\}$ is its *numerical value*, and $[c]$ is the *unit*, so that $c = \{c\}[c] = 299\ 792\ 458$ m/s then the value c is the product of the number $\{c\}$ times the unit $[c]$, and the value never changes. However, the factors $\{c\}$ and $[c]$ may be chosen in different ways such that the product c remains unchanged. • In 1983 it was decided to fix the number $\{c\}$ to be exactly 299 792 458, which then defined the unit of speed $[c]$ = m/s. Since the second, s, was already defined, the effect was to define the metre, m. The number $\{c\}$ in the new definition was chosen so that the magnitude of the unit m/s was unchanged, thereby ensuring continuity between the new and old units.
9	OK, you actually only fix the *numerical value* of the constant expressed in the *new unit*. For the kilogram, for example, you choose to fix the numerical value $\{h\}$ of the Planck constant expressed in the new unit $[h]$ = kg m^2 s^{-1}. But the question remains: suppose a new experiment shortly after you change the definition suggests that you chose a wrong numerical value for $\{h\}$, what then?	• After making the change, the mass of the international prototype of the kilogram (the IPK), which defines the current kilogram, has to be determined by experiment. If we have chosen a "wrong value" it simply means that the new experiment tells us that the mass of the IPK is not exactly 1 kg in the New SI. • Although this situation might seem to be problematic, it would only affect macroscopic mass measurements; the masses of atoms and the values of other constants related to quantum physics would not be affected. But if we were to retain the current definition of the kilogram, we would be continuing the unsatisfactory practice of using a reference constant (i.e. the mass of the IPK) that considerable evidence suggests to be changing with time compared to a true invariant such as the mass of an atom or the Planck constant. [. . .] • The advantage of the new definition would be that we will know that the reference constant used to define the kilogram is a true invariant.

(*Continued*)

Table 5.4 (Continued)

FAQ	Question	Answer
10	Each of the fundamental constants used to define a unit has an uncertainty; its value is not known exactly. But it is proposed to fix its numerical value exactly. How can you do that? What has happened to the uncertainty?	• The present definition of the kilogram fixes the mass of the IPK to be one kilogram exactly with zero uncertainty, $u_r(m_{IPK}) = 0$. The Planck constant is at present experimentally determined, and has a relative standard uncertainty of 4.4 parts in 10^8, $u_r(h) = 4.4 \times 10^{-8}$. • In the new definition the value of h would be known exactly in the new units, with zero uncertainty, $u_r(h) = 0$. But the mass of the IPK would have to be experimentally determined, and it would have a relative uncertainty of about $u_r(m_{IPK}) = 4.4 \times 10^{-8}$. Thus the uncertainty is not lost in the new definition, but it moves to become the uncertainty of the previous reference that is no longer used. . . .
11	The unit of the Planck constant is equal to the unit of action, Js = kg m² s⁻¹. How does fixing the numerical value of the Planck constant define the kilogram?	• Fixing the numerical value of h actually defines the unit of action, J s = kg m² s⁻¹. But if we have already defined the second, s, to fix the numerical value of the caesium hyperfine splitting frequency $\Delta\nu(^{133}Cs)_{hfs}$, and the metre, m, to fix the numerical value of the speed of light in vacuum, c, then fixing the magnitude of the unit kg m² s⁻¹ has the effect of defining the unit kg.
15	Are not the proposed definitions of the base units in the New SI circular definitions, and therefore unsatisfactory?	No, they are not circular. A circular definition is one that makes use of the result of the definition in formulating the definition. The words for the individual definitions of the base units in the New SI specify the *numerical value* of each chosen reference constant to define the corresponding unit, but this does not make use of the result to formulate the definition.

Source: www.bipm.org/en/measurement-units/new-si/faqs.html

In the sections that follow, I discuss aspects of the two fundamental changes to the proposed "New SI" pointed out as (a) and (b) earlier that relate to philosophy of science, especially to the topic of units, quantities, and relations between them. I then address the question of whether the distinction between base units and derived units is still needed after the proposed change to the SI is adopted as an international standard. My answer to this question also explains why we need to retain another distinction that has been questioned in metrology and ignored in much of philosophy of science: the distinction between quantities and dimensions (in the sense of kinds of quantities). I will argue that there is actually a central role for dimensions in the formulation of the SI system, even though they do not appear prominently in the final formulation.

2 What happened to the "base units"?

Consider the form of the first definition of the New SI system of units, shown in Table 5.3 of this chapter (and provided by BIPM to the public in the Draft 9th *SI Brochure* (BIPM, 2013). There are no base units designated. However, if we chose, we could, hypothetically, identify seven quantities as especially selected to be the units defined by the seven "defining constants" of the SI. For example, Hz and m/s are SI units[2] defined by the "defining constant" associated with the hyperfine splitting of Cs, and m/s is an SI unit defined by the speed of light in a vacuum (BIPM, 2013, pp. 12, 29). However, neither of these is a "base" SI unit, in either the current or proposed SI systems.

There are still units designated as the "base units" in the "New SI", though: in terms of proper names, they are the same seven base units of the current SI: metre, kilogram, second, ampere, kelvin, mole and candela. In the first definition of the New SI, they are distinguished in a somewhat indirect, or implicit, way, in that the SI units hertz, joule, coulomb, lumen and watt are related to them (Table 5.3). *Whether using the "explicit-constant" formulation for the seven base units* (Table 5.2), *or the more implicit "defining constant" formulation* (Table 5.3), *the base units will be defined in terms of the exact numerical values that fundamental constants take on when they are expressed in certain (newly defined) SI units.*

Thus, what is at the base of *either* of the definitions of the New SI system of units are seven selected fundamental constants chosen in part for their stability and, thus, their capability to serve as invariant constants of nature. Their numerical values will have zero uncertainty. The reason that the numerical values of the fundamental constants will be exact in the New SI, rather than being the result of measurements containing some uncertainty, is that, in the "New SI", the uncertainty of the constants of nature chosen to define the SI units is zero as a consequence of *fixing* their numerical values with respect to the SI units. That is, *when expressed in the specified New SI units*, their *numerical* value is a certain fixed number. The setting of the numerical value of these constants when expressed in certain SI units effectively defines the SI units. The concisely worded FAQs on the subject developed by the BIPM (BIPM, 2011a) provide more explanation; some selected excerpts are provided in Table 5.4 of this chapter.

This approach involves an unprecedented change in the *general organization of units*: in terms of logical priority, the setting of the numerical value of the fundamental constants is *logically* prior to the definition of the *new, redefined* SI units, since the newly defined SI units are defined in terms of them. The New SI units and the previous SI units will coincide in value *at the moment the change is made*, but there is no expectation that their values according to the current and proposed definitions will continue to coincide after the change, nor that they would have coincided prior to the change. Another subtle detail, which is bound to require some adjustment in pedagogy and in the formulation of investigations about measurement in philosophy of science is that, in the definitions of the New SI, the SI units in which the fundamental constants are expressed when their numerical values are fixed are not limited to the seven base units of the current SI, with which so many are so familiar (although it is still the case that one can express any SI unit in terms of the seven base SI units alone).

What role do the "base units" of the SI play, then, if they are neither at the base of a hierarchy of definitions, nor the only units that figure in the statements fixing the numerical values of the "defining constants" that are at the base of the definitions?

3   Base units in the New SI

As explained earlier, the New SI is striking in that it is the first to propose a definition of the SI system of units that is does not distinguish between base SI units and derived SI units. It does, however, as noted earlier, include an (alternative) definition of the New SI system of units in terms of base units, too, and makes the claim that the two formulations are equivalent. This raises the question of whether the notion of base units is required (Stock & Witt, 2006, p. 586). One thing is not in question: *which* seven units are selected as the base units is not uniquely determined; there is an element of arbitrariness involved (Sterrett, 2009, p. 813; Mohr, 2008, p. 132). However, the point about arbitrariness is distinct from the question of whether the notion of base units can be dispensed with. Mohr has raised the question of the need for the distinction between base units and derived units at all, in light of the fact that one of the definitions of the proposed SI system does not make the distinction (Mohr, 2008, p. 133).

We need to ask, then: if the distinction between base and derived SI units is not essential to providing a definition of the SI system, what are the base units used for, other than organizing the definition of the SI? One answer is that, since the inception of the SI, it seems, the concept of a "coherent system of units" has been characterized in terms of base units. When Burdun (1960) explicitly listed the provision of a "coherent system of units" as a requirement of a system of units, he clarified what he meant, and did so in terms of "basic units" and "derived units": "In selecting the basic units of the system it is necessary: 1) to provide a coherent system of units, i.e., to select basic units which would produce derived units by multiplication or division without introducing numerical coefficients" (Burdun, 1960, pp. 913–914). Thus, if base units were not specified, this account of the coherence of the SI system of units could not be applied to produce a coherent system of units, or to verify that a certain system of units was coherent.

Yet one of the points celebrated about the New SI is that the definition of the SI can now be made without mentioning base units or derived units at all (Bordé, 2005; Mohr, 2008). The way I see to make sense of this is as follows: the "more fundamental" definition of the SI, the one made in terms of the seven "defining constants" without mentioning base units (Table 5.3), is one way to define the SI, and that definition can be used to identify *specific quantities* for the SI units (some base, some derived) that are mentioned in that definition. This is just what the definitions referring to artefacts such as the standard metre and the standard kilogram did for the base SI units (as Mohr also points out in Mohr, 2008, p. 133), but this definition does so without mentioning any artefacts. This more fundamental definition also ties all the SI units that it does mention to the units that *happen to be the units previously identified as SI base units*, thus allowing the question of which units are coherent SI units to be answered in the same way as before. Thus, the question of coherence of units of the "New SI" is answered, without having to be addressed anew.

It seems to me the New SI raises, in turn, a foundational question about *coherence of a system of units*. We have seen that there is a *practical value* in having the distinction between base units and derived units in place when it comes to specifying which units of the SI are coherent SI units: doing so makes it fairly simple to define a coherent SI unit, and, accordingly, simplifies confirming that a unit is a coherent SI unit. Does coherence of a system of units *require* the distinction, though? Raising this question leads in turn to the question as to what the coherence of a system of units consists of, and why it is considered essential to a system of units.

4 Coherence of a system of units

In a previous article I used a different characterization of the *coherence of a system of units* than Burdun gives, and one I think not uncommon, i.e., that "[a] system of units is coherent if the relations between the units used for the quantities is the same as the relation between the quantities in the fundamental equations of the science" (Sterrett, 2009, p. 806); the title of this chapter comes from that characterization. In laying out the topic there, I found it important to discuss not only units and quantities, but also "dimensions" (sometimes called "kinds of quantities"). Although in that article I referred only to the SI system and associated documents in place at that time (the current SI), many of the major points are still relevant, since the New SI likewise uses dimensions in its formulation of the distinction between base units and derived units (Draft 9th *SI Brochure*, section 1.3).

The usefulness and necessity of the concept of dimension has been questioned (e.g., Emerson, 2005). However, I will argue, the points that have been raised as reasons for questioning the need for the concept of dimension are mistaken, largely because they do not take into consideration the significance of the distinction between quantities and dimensions. In addressing the topic of whether the distinction between base units and derived units is needed, we shall see how important and useful the concept of dimension is, once we get clear about the distinction between "quantity" and "dimension" (a distinction which is made in both the current SI and the New SI).

More recently, another relevant article about the coherence of a system of units has appeared (de Courtenay, 2015). The two articles (my "Similarity and dimensional analysis" (Sterrett, 2009) and de Courtenay's "The double interpretation of the equations of physics and the quest for common meanings" (de Courtenay, 2015)) are concerned with different aspects of the topic, but are complementary. Put briefly, whereas I was concerned with drawing out the consequences of what is built into a system of units that meets the constraint of being a coherent system of units, de Courtenay's article explains how that constraint is actually achieved. She also discusses why the constraint has been so important to modern science, what it conceals, and the situation we are left in even after it has been achieved. I thus refer to some points made by de Courtenay in "The double interpretation of the equations of physics and the quest for common meanings" (de Courtenay, 2015) as well as those I have made in my "Similarity and dimensional analysis" (Sterrett, 2009).

4.1   Units, quantities and dimensions

I will use the characterization that a system of units is *coherent* if the relations between the units used for the quantities are the same as the relations between the quantities in the fundamental equations of the science (Sterrett, 2009, p. 806). It follows from this that if one is concerned with making sure that a certain system of units being formulated is to be coherent, then, they would need to know the relations between quantities prior to formulating a system of units. This would require being able to write the fundamental equations of science in a way that is *independent of* a system of units. One might well ask if this is indeed possible?

The answer is that, yes, it is possible to write the fundamental equations of science in a way that is independent of a system of units, if the fundamental equations are regarded as *quantity equations.* In Sterrett (2009), I discussed quantity equations, contrasting Lodge's recognition of the fundamental equations of mechanics and physics as quantity equations with James Clerk Maxwell's view of equations. On Maxwell's view of equations, a physical quantity appearing in a physical equation must be measurable, which means that the value of a quantity consists of two parts: a numerical part, and a unit with which all quantities of that kind can be compared. As I noted there:

> Maxwell explicitly discussed this conception of a quantity, while recognizing ambiguities in the notation of physical quantities as used in equations in scientific practice. He noted that symbols used as variables in equations of physics lent themselves to two different interpretations: (i) as denoting the lines, masses, times, and so on themselves, and (ii) "as denoting only the numerical value of the corresponding quantity, the concrete unit to which it is referred being tacitly understood".[3]
>
> Each of these interpretations presents a problem, though. . . . The first interpretation doesn't really apply during the process of performing the

numerical calculations. . . . The second interpretation doesn't satisfy the requirement that "every term [of an equation of physics] has to be interpreted in a physical sense." Maxwell's way of resolving the ambiguity he identified was to take a sort of hybrid approach.

(Sterrett, 2009, p. 804)

Alfred Lodge's approach aimed to avoid the ambiguity that Maxwell identified (what de Courtenay (2015) calls "the double interpretation of the equations of physics") by regarding the equations of mechanics and physics as equations that "express relations among quantities" (Lodge, 1888, pp. 281–283), i.e., as *quantity equations*. This is the approach I took in Sterrett (2009), in which I identified the feature of *coherence of a system of units* as the key to understanding why it is possible to use dimensionless parameters to establish a similarity of physically similar systems. For my purposes in that paper, I, too, could avoid the ambiguity in the equations of physics that Maxwell identified, as Lodge did, by regarding the fundamental equations of mechanics and physics as quantity equations. Lodge pointed out that, understood as quantity equations, the fundamental equations of a science

are independent of the mode of measurement of such quantities; much as one may say that two lengths are equal without inquiring whether they are going to be measured in feet or metres; and, indeed, even though one may be measured in feet and the other in metres.

(Lodge, 1888, pp. 281–283)

Guggenheim, a student of Lodge's, wrote a piece intended as a reference for others on the topic called "Units and Dimensions" (Guggenheim, 1942). Guggenheim there states his view that "we are entitled to multiply together any two entities, provided our definition of multiplication is self-consistent and obeys the associative and distributive laws". He argues that it is "perfectly legitimate to multiply together any two physical entities, such as a length and a force. If the reader naïvely asks: 'What, then, is the product of a foot and a pound?' I reply a 'foot-pound'" (Guggenheim, 1942). That view is now fairly well accepted (de Boer, 1995; VIM, 2012). The VIM (Vocabulary of International Metrology) includes the term "quantity equation", giving the meaning as "mathematical relation between quantities in a given system of quantities, independent of measurement units" (VIM, 2012, §1.22). De Boer explains the significance of a relation between quantities that is independent of measurement units, following up on Maxwell's arguments:

Expressing the results of physics in terms of physical quantities has the advantage of giving a representation which does not depend on the choice of unit. When a particular length L is expressed in two different units [L]' and [L]" by the two expressions $L = \{L\}' \times [L]'$ and $L = \{L\}'' \times [L]'' \ldots$ the *physical quantity* L itself is an invariant [as Maxwell argued.] This is an important argument and an essential reason in favour of using quantities and not numerical values in the theoretical description of physical phenomena; *Using physical quantities gives a representation which is invariant with respect to the choice of units.*

(de Boer, 1995, p. 406)

This approach was used in discussions about the transition from CGS units to the Giorgi/SI units. Cornelius remarked in 1964 that using quantity equations led to a dispute as to whether the difference between the two systems lay in different units or different quantities. The answer, of course, was that the "radical" change was a change in the form of the equations – which led to a change in the coherence of units (Cornelius et al., 1964, p. 1446). The terms "quantity calculus" and "quantity equation" are still used. Very recently, in discussing the possible roles the international prototype of the kilogram might still play after the proposed revisions to the SI are adopted, Davis likewise found it useful to do so (2011), citing de Boer's "On the History of the Quantity Calculus and the International System" (1995).

Dr. de Courtenay, in contrast, faces the ambiguity head-on, rather than avoiding it. She shows Maxwell's view on equations to be something that is "now taken for granted by every student of physics when solving a problem and sliding from the algebraical solution to its numerical application" (2015, pp. 145–146). Maxwell's view of the equations of physics serves as the starting point of her inquiry into the double interpretation of equations; she proceeds to discuss how terminology has since evolved and explains why in fact it is now appropriate to regard the "double interpretation" view as cogent:

> The smooth passage between the two interpretations was achieved by the construction of a *coherent* system of units which ensured that the equations had *exactly the same expression under both interpretations.* The coherent system of units, which was to be disseminated worldwide by the metrological international organizations, reconciled the two formerly conflicting points of view within a framework that warranted, at last, the double interpretation of the equations of physics propounded by Maxwell.

(de Courtenay, 2015, p. 146)

The process involved in establishing a coherent system of units that proceeds starting from relations that are invariant under a change of units is laid out in de Courtenay (2015, pp. 146–147). These are "accepted physical relationships that can be stated in relations of proportions", and choices at this initial stage "determine the *structure* of the system" (p. 146). Dr. de Courtenay's account of what goes into producing a coherent system of units provides valuable background for a better understanding of the point in my previous paper. I had emphasized that coherence of a system of units is relative to the quantity equations one chooses to use, but I did so based upon looking at an historical episode, rather than providing a logical analysis of the process:

At one time, the CGS system provided a coherent set of units for Newtonian mechanics, yet it was not clear what to say at that time about a coherent set of units for electromagnetism. There was a CGS system for Newtonian mechanics, a CGS-M system of units for magnetism, and a CGS-E system of units for electrical phenomena. It was shown in 1901 by Giorgi that the CGS system could be amended in alternate ways to provide a coherent set of units for the quantities in the equations describing electromagnetic phenomena, so that a choice had to be made. The decision made by the committees governing the SI system was to include a base unit for the physical quantity of electrical current; the SI system now provides a coherent set of units with respect to Maxwell's equations as well as for Newton's. When used for electromagnetism, the SI system (which follows Giorgi) and the older CGS systems are really different systems of units; for electromagnetic phenomena, unlike for mechanics, switching from one of these systems to the other involves more than a simple change of units, for there will be some equations whose form differs depending on which system one is using. Thus, some of the units for the older CGS systems are referred to as non-SI units.

(Sterrett, 2009, p. 809)

The question being investigated there was: what underlies inferences based upon dimensionless parameters, or more precisely, based upon the identity of the values of dimensionless parameters between two different systems? The answer was that, once a coherent system of units is in place, the fact that one is using a coherent system of units is relied upon (whether or not it is explicitly recognized), when using the numerical values that dimensionless parameters take on to establish similarity of physical systems (Sterrett, 2009, 2017a, 2017b). That a ratio of quantities is dimensionless is significant because it says something about the relations between the quantities in it. The underlying logic involves dimensions as well as quantities and units, and further examination will reveal a role for dimensions in drawing a distinction between base units and derived units.

Dimensions are to be distinguished from units and from quantities, although they are frequently confused with both of them, in print as well as in discussion, sometimes even in textbooks. Dimensions have to do with a constraint on equations, a constraint which we know the equations of physics, if we've gotten them right, must be subject to. Dimensions can be used to formulate a necessary and sufficient criterion for an equation to be homogeneous – even an equation containing undetermined quantities. The requirement that an equation be homogeneous is like the requirement that a sentence be grammatically correct; just as a sentence is not a proper sentence unless it is grammatically correct, so an equation is not a physical equation unless it is dimensionally homogeneous. Now, the question often arises as to whether one might just as well appeal to units for this purpose, and hence, whether it is possible to regard dimensions as superfluous (i.e., whether it is possible to hold that dimensions can be dispensed with, or, alternatively, to hold that dimensions just are units). Although this is often done, we shall see that things are not so easy, and that a lack of clarity about both units and dimensions is very lamentable and may have contributed to an ill-advised disregard for an appreciation of the role of dimensions.

Dimensions appear in the New SI (Draft 9th *SI Brochure*) just as they do in the current SI (*SI Brochure* 8th edition, 2014 update); they are referred to as "dimensions of quantities". Each SI base unit has a unique dimension associated with it; the dimensions of other quantities are constructed from products of powers of these dimensions. The *SI Brochure* indicates as a rationale for providing dimensions only that "By convention physical quantities are organised in a system of dimensions" (BIPM, 2013, section 1.2, p. 3). This does not help much in understanding what dimensions are, nor in determining whether the fact that each base unit is associated with a unique dimension means that defining some units as base units is essential in the SI system. In order to understand what dimensions are, we go back to historical roots of the concept. I find Fourier's comments on dimensions clearer and more reliable than Maxwell's. And, since Maxwell credits Fourier as the first to state the theory of the dimensions of physical quantities, I take it that Maxwell means to be endorsing the points about dimension found in Fourier's *Analytical Theory of Heat*. How Fourier speaks about dimensions, and what he says about them, is helpful in clarifying the distinctions between dimensions, quantities and units, and will lead us to the topic of base units.

Fourier speaks in terms of *exponents of dimension*, and he speaks of the dimension of (what we would call) a variable in the equation *with respect to* a unit, i.e., "the dimension of x *with respect to* the unit of *length*". For instance, a number representing a surface has dimension 2 (with respect to the unit of *length*), and one representing a solid has dimension 3 (with respect to the unit of *length*). His discussion focuses on equations; an equation expresses a relation between magnitudes. He notes two things: (a) the terms of an equation cannot be compared unless they have the same exponent of dimension; and (b) "every undetermined magnitude or constant has one dimension proper to itself" (Fourier, 1878, section 160, p. 129). The idea in (b) is that one can *deduce the dimensions* of an undetermined magnitude occurring in an equation from these two principles. That is, the *equation* expresses how the magnitudes and constants occurring in it are related to each other; and the *dimensions* of each magnitude or constant indicate how each of them is related to each of the units (for each "kind of unit"). Then, the requirement of homogeneity puts a collective constraint on the dimensions of the magnitudes and constants that occur in the equation.

Fourier identifies three units as being relevant in his analysis of equations in the theory of heat: length, duration and temperature; these are independent of each other. (That they are independent on Fourier's account is clear from a table he presents in which he shows that the dimension of the magnitude he calls x with respect to length is 1, with respect to duration is 0, and with respect to temperature is 0. If we express this by saying that the dimension of l (length) is 1,0,0, we can express the situation with the others by saying that the dimension of t (duration) is 0,1,0; and that of v (temperature) is 0,0,1.) Thus this collective constraint on the dimensions of the magnitudes and constants that occur in an equation in the theory of heat is actually the *conjunction of three separate constraints on the equation*: the dimension of every term of the equation with respect to the unit of length must be equal, *and* the dimension of every term of the equation with respect to unit of duration must be equal, *and* the dimension of every term of the equation with respect to the unit of temperature must be equal.

Now, in saying that (on Fourier's account) the dimensions of a certain magnitude or constant indicate how that magnitude or constant is related to each of the units he identifies, I do not mean that in order to carry out the dimensional analysis, one must refer to, or assume, any *particular* unit or system of units. Fourier's starting point is an equation relating magnitudes and constants, and, as he repeatedly emphasizes, such an equation is the same no matter what units one is using. The kind of equation he is talking about is in fact invariant with respect to changes of units, but the kind of changes he means are only changes that are a matter of size of the unit. Fourier works out an example to illustrate that, although individual magnitudes in the equation will change to accommodate a change in (the size of) units, they will change in unison in such a way that the equation itself holds without modification (Fourier, 1878, pp. 128–130). Thus, the constraint of dimensional homogeneity on the equation involves reference to the existence of some units, but not any particular units, and the constraint itself is put in terms of exponents that refer to units very generally, i.e., as "unit of length", "unit of duration" and "unit of temperature".

To get clear on the respective roles of units and dimensions: on Fourier's account, the notion of dimension draws on the rather minimal assumption of the existence of a system of units in which the magnitudes and constants in the equation can be measured (or, perhaps just the possibility of the existence of one), but not on the existence of any particular choice of units. What features of a system of units is he relying on, though? As mentioned earlier, the three units Fourier uses in analysing his equation (one unit each for length, duration and temperature) are independent of each other. Does his analysis assume that *any* systems of units one might use has the feature that there is a set of independent units, too? What role do dimensions have in his reasoning?

As I will show, in Fourier's analysis (as in the SI brochures for both the current SI and the New SI), dimensions are not the same thing as units. We can see that there is an intimate relation between them, though, in spite of the fact that they are not the same thing, by reflecting upon Fourier's analysis of (dimensional) homogeneity of equations. Recall that Fourier speaks of the dimension of a magnitude *with respect to* a unit of a certain quantity for a particular magnitude. Thus, *dimensions cannot be the same thing as units* because dimensions stay the same throughout many changes in size of units. For example, in Fourier's analysis, as he exhibits in the form of a table, the dimension of the quantity "surface conducibility *h*" with respect to length is –2 (Fourier, 1878, section 161, p. 130), and the exponent of the dimension will stay the same for any size of unit for length one chooses. The exponent of dimension of a given magnitude with respect to the unit of length thus implicitly assumes the *possibility of a unit for length* and indicates how each of the magnitudes involved in an equation is related to the unit for length.

Yet, in general, it is not a requirement of a system of units that it contain a unit for length, nor for any particular quantity. Fourier's analysis is thus not an analysis of the most general case, for it assumes a system of units with some special features, i.e., that the system contains units for the magnitudes of length, duration and temperature. (He mentioned that a total of five were needed for physics.) Perhaps that assumption was not inappropriate for his purposes. However, we are interested here in a more general question. Fourier's formulation is very useful in that it lends itself to generalization: he spoke of the exponents of the dimensions of the unit for length, for duration and for temperature. This way of speaking does allow us to generalize away from a specific set of units, by talking about "unit for length", "unit for duration" and "unit for temperature". What are the things that we wish to generalize over (length, duration, temperature) called then, if not units? They might be called kinds of quantities or magnitudes, but they are distinguished among other quantities in that they are selected as belonging to a small group of kinds of quantities or magnitudes that are of some special relevance to the kind of equation being analysed. We already saw that for Fourier, they are independent of each other.

As Fourier lays out his analysis of the equations of the theory of heat in terms of the exponents of the dimension of each of three selected magnitudes, the formalization he uses provides a canonical format for associating exponents of dimensions with each magnitude or constant that might occur in an equation in the theory of heat. Using the term dimension to generalize, we would then say that Fourier used the three *dimensions* for length, duration and temperature in his analysis, and assumed that units for each of the three quantities of length, duration and temperature could be identified, although he did not refer to any specific system of units. Such a formulation permits generalizing from those three specific dimensions. We might put it this way: the role that the dimensions for units of length, duration and temperature have is to provide a logical tool for implementing the grammatical constraint of dimensional homogeneity on the equation, independently of reference to a system of units. A dimensional equation consists of sums of products of dimensions with exponents; it is independent of the choice of units one uses, and so is invariant with respect to changes in (sizes of) units. It is very easy to blur things here and conflate dimensions with units or quantities. Many have identified dimensions with either units or quantities, as a matter of convenience. It may be there are some contexts in which that does no harm. However, I will argue for the importance of distinguishing dimension from either units or quantities in foundational or methodological investigations.

Since the dimensions of a quantity do not always determine which quantity it is, dimensions are not quantities. Lodge made a point of this, too, noting that *work* and *moment of force* are not the same quantity, but they have the same dimensions of quantity. The point bears emphasis: there are quantities we consider distinct which, when subjected to the kind of analysis Fourier described, yield the same exponents of dimension. This is not a reduction of the notion of dimension, however, as Emerson's comments appear to imply (2005). Rather, these examples merely illustrate the basic fact that dimensions are not the same thing as quantities; they show nothing regarding the uselessness of dimensions. Dimensions have an essential role to play in formulating the constraint of dimensional homogeneity on equations. For this purpose, a few quantities determined to be independent are designated as base dimensions. The set of independent base dimensions selected for the analysis then provides a canonical format for expressing the exponent of dimensions for any quantity, just as a set of selected vectors provides a canonical format for expressing every vector, or factorization by prime numbers provides a canonical format for expressing every positive integer. If by analogy with base (or basis) vectors, we call this set of independent dimensions the base dimensions, it is then straightforward to write an equation for a given dimension that expresses the constraint that all the terms in the equation have the same exponent of (that particular) dimension. We can write such an equation for each of the dimensions. A dimension is thus associated with a quantity: for each dimension, a certain quantity is selected to serve the role that length, duration and temperature (which Fourier called a "kind of unit") served in Fourier's analysis when he spoke of the exponents of dimensions of magnitudes and constants *with respect to* length, duration and temperature. It is true that, as Fourier puts it, each magnitude or constant in an equation [of the theory of heat] has "one dimension proper to itself", a fact that he regards as derived from "primary notions" about quantity (Fourier, 1878, p. 128). But, as we've seen, this does not mean that there is a single quantity associated with each dimension; that point is entirely consistent with the existence of two different quantities that have the same dimensions (Sterrett, 2009; Lodge, 1888). As I have written on an earlier occasion:

> One of the principles used for the [current] SI System is that the dimension of every quantity, whether base or derived, is unique; that is, there is only one such dimension of canonical form associated with each quantity Q. Since the number of derived quantities is unlimited, and the number of dimensions of canonical form is unlimited [as well], one may ask whether there is a unique quantity associated with each dimension of canonical form. The answer is no: more than one quantity may have a given dimension associated with it, just as more than one quantity may have the same units (*heat capacity* and *entropy* are considered physically distinct quantities, though they are both measured in joule/Kelvin; *electric current* and *magnetomotive force* are both measured in amperes).
>
> Thus, one cannot infer from a dimension, the quantity with which that dimension is associated, for the quantity is not uniquely determined.
>
> (Sterrett, 2009, p. 813)

To summarize: the *concept of dimension* is important to preserve, because of the role it has in the criterion for the homogeneity of equations in physics. Whereas, the *concept of quantity* is important because of the role that quantities have in the quantity equations expressing the fundamental laws of science.

4.2   Quantity equations and dimensional equations

On Lodge's approach (as with the approach I took in Sterrett (2009)), quantity equations express the fundamental laws of physics as relations between quantities. Quantity equations must be dimensionally homogeneous, but the constraint of dimensional homogeneity on an equation is like a grammatical constraint on a statement; just as there is much more to a statement than its grammatical structure, so there is much more to an equation of physics than homogeneity of dimensions. Another way to put this is to point out that the dimensional equation one would write to show that the quantity equation is dimensionally homogenous is not necessarily as informative as the quantity equation itself is. For example, consider the equation $s = a\,t^2$, where s is a distance, a is an acceleration and t is a time, considered as a quantity equation. The dimensional equation is $[L] = [L][T]^{-2}[T]^2$ or $[L] = [L]$.

Where do dimensional equations fit in the logical scheme of developing a system of units? They are not the beginning point, for dimensional equations are *relative* to the quantity equations expressing the fundamental laws of the science one is concerned with. (We saw that, even in Fourier's presentation of how to determine the exponents of dimensions of the units for various quantities, a necessary prior step was identifying the three quantities with respect to which the exponents of dimension were to be determined.) Dimensional equations are also relative to the selection of *which* of the quantities in those quantity equations is designated for use as a *base dimension* in the dimensional analysis.

Thus, in a rational reconstruction of dimensional equations, the logical order of things seems to be (a) identifying the quantity equations expressing relations between quantities; (b) showing that there can be a system of units for the quantities associated with those quantity equations; (c) selecting the quantities upon which the dimensional analysis yielding the dimensional equation is to be based; and (d) writing the dimensional equations that result from performing a dimensional analysis of a quantity equation one is interested in. The specific dimensional equation that results will depend upon the choices in (a) and (c).

4.3   Constructing systems of units

Now, the rational reconstruction of dimensional equations just discussed does not include developing a coherent system of units. There is another step, the one de Courtenay calls the final and most crucial step:

> Only in the end do the units of the base magnitudes, called base units, get selected. The units of the derived magnitudes are then fixed as derived units

expressed in terms of the base units through the structure, or the other equations, without introducing any numerical coefficient (i.e., by setting all the coefficients equal to one). It is this last step that is all important: it established the bridge between the two interpretations and at the same time conceals the gap between them because it ensures that the equations interpreted in terms of magnitudes and in terms of measures have exactly the same expression.

(de Courtenay, 2015, pp. 146–147)

However, in later sections of the paper, she points out that the gap still exists, in that "the mathematical equations of physics, in their theoretical use, are primarily to be understood as equations between magnitudes", which is very much like understanding them as quantity equations. Her explanation makes it clear that achieving the goal of a coherent system of units involves activities prior to this last step: identifying the quantity equations (from observation, experiment and analysis), and choosing which magnitudes to designate as base magnitudes. This is meant to describe the current and past SI systems. These reflections reinforce the value of "thinking in terms of requirements placed on a *system* of units, rather than in terms of requirements placed on the units chosen for each kind of quantity individually", which was also helpful in understanding the historical case of the change from CGS units to the Giorgi/SI units I discussed in Sterrett (2009). In that case, which was a revision of a system rather than the construction of a new one from scratch, thinking in terms of coherence of a *system* of units rather than about the change to individual units led to understanding that a change in the form of the quantity equations could bring about a change in the coherence of units (Sterrett, 2009, p. 809).

For any coherent system of units, then, the establishment of the system of units, including the definitions of the units, is a holistic affair that starts with the quantity equations. Since it is part and parcel of the International System of Units that it be a coherent system of units, really, the units in most earlier versions of the SI are in some sense collectively defined. Yet there is a difference between the New SI and its predecessors due to the definitions no longer including any artefacts in the New SI: now that no base unit is defined in terms of an artefact, that reason for defining base units is gone. The other important change effected by the New SI, the decoupling of the definition and the "practical realization" of the units, removed another reason one might need to designate some of the SI units as base units. There was another objective that did call for designating some of the units as base units, and that reason did not go away – however, it is parasitic in the fact that the previous systems employed a distinction between base units and derived units: the objective of keeping the values of the base quantities continuous at the moment of transition from the use of the current SI to the use of the New SI.

However, referring to de Courtenay's description of what is involved in establishing a coherent set of units, we see that even though the designation of base units does not occur until the last step, the designation of base *magnitudes* occurs in a previous step. The "last step" of the process taken to conceal the gap between what she calls magnitude equations and measure equations does not involve a free choice as to which units to designate as base units. Rather, on her account, the designation of base units is determined by the prior choice of which magnitudes were selected to be base magnitudes, since base units are "the units of the base magnitudes" (de Courtenay, 2015, p. 147). Thus, the "magnitude equations" and choice of base magnitudes are really what organize the base units. Then, the relation between the derived and base units is structured as needed to accord with the magnitude equations.

As I employ Lodge's idea of quantity equations rather than Maxwell's "double interpretation", my discussion is slightly different, but its bottom line is analogous: prior to the selection of base units, one possible designation of which units to call the base units is already available, due to the choice of *base dimensions*, which is logically prior. I do not see, however, that the choice of base units is constrained by this. (According to the Draft 9th *SI Brochure*: "The choice of the base units was never unique" (BIPM, 2013, section 1.3).) By selecting the units of the base dimensions to be the base units, though, the rest of the process is made much easier than by making any other selection for the base units. *The dimensions help organize the units by providing a formalism in which relations between quantities, with respect to a set of quantity equations expressing the fundamental relations of a science, can be more perspicuously studied. Dimensions, however, do not appear prominently in the formulation of the system of units, once it has been constructed.*

5   Conclusion

We are now ready to address the question of whether there is a need for the distinction between base quantities and derived quantities in a system of coherent units, as opposed to abandoning the distinction. The answer is both a qualified yes and a qualified no.

The answer to the question of whether the distinction is needed is (a qualified) yes. Obviously, it is possible to construct a definition of the SI that does not designate any units as base units, as evidenced by the "defining constants", or "single scaling statement" formulation of the New SI. Hence it may seem unequivocal that it is possible to define a coherent system of units without designating some of them as base units. However, it seems to me that some distinction between base and derived units necessarily occurs in the logical underpinnings of a coherent system of units, in that a choice of base units, even if only a provisional one, has to be made in the steps taken prior to the definition of units. This is because the formalism developed to understand how quantities are related to each other – the concept of exponents of dimension, or dimensional analysis – does involve the notion of base units. (Recall that Fourier used "the unit for length", "the unit for duration" and "the unit for temperature" as base units in performing a dimensional analysis.) This choice of base units, which is needed to carry out the dimensional analysis of quantity equations in order to determine the dimensions of quantities, is logically prior to defining the (value of the) units. There is some freedom of choice in selecting which dimensions to designate as base dimensions and, from a logical standpoint, one need not designate any particular dimension as a base dimension. In fact, in the New SI, on the "definitional constants", or "single scaling statement" definition, there does not even seem to be a *logical* constraint anymore that would prevent someone from choosing to designate one set of units as base units when using the New SI for one application, and then to designate another set of base units for another problem or situation. There are, of course, often practical constraints and disadvantages associated with doing so.

It is worthwhile reflecting on why this "yes" answer is so. Because the starting point for formulating a coherent system of units is a collection of quantity equations that expresses relations between quantities, the relations between any quantity included in these equations and the selected base quantities will be available, even if not explicitly stated. The quantity equations are actually discussed in the *SI Brochure*, which says that

> The system of quantities used with the SI, including the equations relating the quantities, is just the set of quantities and equations that are familiar to all scientists, technologists, and engineers. They are listed in many textbooks and in many references, but any such list can only be a selection of the possible quantities and equations, which is without limit.

There are various international standards cited, but there is now a single standard comprising them: "the ISO/IEC 80000 Standards, Quantities and Units, in which the corresponding quantities and equations are described as the International System of Quantities". A telling detail is found in a footnote:

> In these equations the electric constant ε_0 (the permittivity of vacuum) and the magnetic constant μ_0 (the permeability of vacuum) have dimensions and values such that $\varepsilon_0\mu_0 c^2 = 1$, where c is the speed of light in a vacuum. Note that the electromagnetic equations in the CGS-EMU, CGS-ESU and Gaussian systems are based on a different set of quantities and equations in which the magnetic constant and electrical constant have different dimensions, and may be dimensionless.
>
> (BIPM, 2013, p. 2)

This note illustrates an important point about the history of the SI: the quantity equations and system of quantities developed for these standards already incorporate a concern for being suitable for use in developing a coherent a system of units. This just reflects the fact that the development of coherent systems of units was a holistic affair that included the quantity equations; sometimes the logical order of things even ran backwards, from an already established unit in use to reformulation of quantity equations to accommodate using it. Using the quantity equations that have been developed in these standards, no matter what units are designated as base units, the relations between any SI unit and the selected base units will be deducible; the relations between the base units and derived SI units do not have to be *created* by another statement of definition that stipulates how they are related. The theory of dimensions can be used to articulate what those relations are. A remark in the *SI Brochure* puts things slightly different, though I think it reflects the same point:

> The dimension of a derived quantity provides the same information about the relation of that quantity to the base quantities as is provided by the SI unit of the derived quantity as a product of powers of the SI base units.
>
> (BIPM, 2013, p. 2)

The answer to the question (as to whether there is a need for the distinction between base quantities and derived quantities in a system of coherent units) *is a qualified no* in the following sense: a definition of the units of the SI that distinguishes some of the units as base units is no longer needed, in the sense of requiring an *additional* step of selecting base units, defining them, and stating the relations between SI units and the base units.

This is how the issue appears to me, for the reasons stated in this paper. More importantly, the two striking features of the New SI that I identified near the outset of this chapter as features I expect many philosophers of science to find surprising present an opportunity. These features were that

> the proposed system of international units ('New SI') can be defined (a) without drawing a distinction between base units and derived units, and (b) without restricting (or, even, specifying) the means by which the value of the quantity associated with the unit is to be established.

These challenge naïve notions about measurement, such as that values of units are tied to artefacts, or that definitions of quantities are tied to specifically defined processes, which one often finds in philosophical discussions of measurement. On the one hand, as many have noted, the New SI is the realization of a vision that physicists of the nineteenth century had outlined, and so, in some sense, it is not new. On the other hand, however, this is the first time we have a system of units realizing that vision. It is an occasion for philosophers of science to re-examine presumptions about measurement and the logic of the equations of physics. I also hope that my study in this chapter will encourage an appreciation of what I feel is an essential part of the language of science: the theory of dimensions.

Acknowledgments

Thanks to Nadine de Courtenay for discussion, and to two anonymous reviewers for comments on an earlier version of this chapter.

Notes

1 Paper submitted 2015; Tables 5.2 and 5.3 updated December 2018.
2 The quantity velocity is measured in m/s, which is identified as one of the SI derived units without a special name. (Other derived SI units without a special name are m^3, m^2 and m/s^2.)
3 Maxwell (1890, p. 241).

References

BIPM. (2011a). *FAQs, frequently asked questions about the New SI*. Paris: Bureau International des Poids et Mesures. Revised October 2011, retrieved 28 June 2015 from www.bipm.org/en/measurement-units/new-si/faqs.html.

BIPM. (2011b). *Future revision of the SI: Explicit-constant formulation [of units]*. Paris: Bureau International des Poids et Mesures. Retrieved 28 June 2015 from www.bipm.org/en/measurement-units/new-si/explicit-constant.html.

BIPM. (2013). *Draft 9th SI Brochure*. Paris: Bureau International des Poids et Mesures.

BIPM. (2014). *"Resolution 1" in Resolutions adopted by the CGPM at its 25th meeting, 18–20 November 2014*. Paris: Bureau International des Poids et Mesures. Retrieved from www.bipm.org/en/CGPM/db/25/1/.

BIPM. (CCE-09-05). *Mise en pratique CCE-09-05*. Paris: Bureau International des Poids et Mesures. Retrieved from www.bipm.org/cc/CCEM/Allowed/26/CCEM-09-05.pdf.

BIPM. (undated). *"What" tab of "On the future revision of the SI"*. Paris: Bureau International des Poids et Mesures. Retrieved 12 December 2017 from www.bipm.org/en/measurement-units/new-si/.Bordé, C. (2005). Base units of the SI, fundamental constants and modern quantum physics. *Philosophical Transactions of the Royal Society A: Mathematical, Physical and Engineering Sciences*, 363, 2177–2202, 2182.

Bordé, C. (2018). Reforming the international system of units: On our way to redefine the base units solely from fundamental constants and beyond. [This volume].

Burdun, G. D. (1960). International system of units. *Measurement Techniques*, 3, 913–919.

Cornelius, P., de Groot, W., & Vermeulen, R. (1964). Quantity equations and system variation in electricity. *Physica*, 30, 1446–1452.

Davis, R. S. (2011). The international prototype of the kilogram after redefinition of the International System of Units. *Philosophical Transactions of the Royal Society A: Mathematical, Physical and Engineering Sciences*, 369, 3975–3992.

de Boer, J. (1995). On the history of quantity calculus and the International System. *Metrologia*, 31, 405–429.

de Courtenay, N. (2015). The double interpretation of the equations of physics. In O. Schlaudt & L. Huber (Eds.), *Standardization in measurement: Philosophical, historical and sociological issues*. London: Pickering & Chatto.

Emerson, W. H. (2005). On the concept of dimension. *Metrologia*, 42, L21–L22.

Fourier, J. (1878). *The analytical theory of heat* (A. Freeman, Trans.). Cambridge: Cambridge University Press.

Guggenheim, E. A. (1942). Units and dimensions. *The Philosophical Magazine*, 33, 479–496.

Lodge, A. (1888). The multiplication and division of concrete quantities. *Nature*, *38*, 281–283.

Maxwell, J. C. (1890) [1877]. Dimensions. In *The Encyclopaedia Britannica: A dictionary of arts, sciences, and general literature* (pp. 240–242), Ninth Edition, Volume VII. New York, NY: Henry G. Allen Company.

Mills, I. M., Mohr, P. J., Quinn, T. J., Taylor, B. N., & Williams, E. R. (2011). Adapting the International System of Units to the twenty-first century. *Philosophical Transactions of the Royal Society A: Mathematical, Physical and Engineering Sciences*, *369*, 3907–3924.

Mohr, P. J. (2008). Defining units in the quantum based SI. *Metrologia*, *45*, 129–143.

Quinn, T. (2018). The origins of the Metre Convention, the SI, and the development of modern metrology. [This volume]

Riordan, S. (2015). The objectivity of scientific measures. *Studies in History and Philosophy of Science*, *50*, 38–47.

Sterrett, S. G. (2009). Similarity and dimensional analysis. In A. W. M. Meijers (Ed.), *Handbook of the philosophy of science. Volume 9: Philosophy of technology and engineering sciences* (pp. 799–823). Amsterdam: North Holland.

Sterrett, S. G. (2017a). Physically similar systems: A history of the concept. In L. Magnani & T. Bertolotti (Eds.), *Springer handbook of model-based science* (pp. 377–411). Heidelberg: Springer.

Sterrett, S. G. (2017b). Experimentation on analogue models. In L. Magnani & T. Bertolotti (Eds.), *Springer handbook of model-based science* (pp. 857–878). Heidelberg: Springer.

Stock, M., & Witt, T. J. (2006). CPEM 2006 round table discussion 'Proposed changes to the SI'. *Metrologia*, *43*, 583–587.

VIM. (2012). *Vocabulary of international metrology* (3rd ed.). Retrieved from http://jcgm.bipm.org/vim/en/1.22.html.

6 On the conceptual nature of the physical constants[1]

Jean-Marc Lévy-Leblond

Introduction

In most formulae of physics, or, more generally, in most theoretical analyses of any physical phenomenon, there appears one or more physical constants. Some of these play an essential and pervasive role in physics. They are variously called "general", or "fundamental", or "universal" physical constants. Yet, despite their importance, very little seems to have been written about their nature and significance until rather recently (save for Duff, Okun, & Veneziano, 2002; Uzan & Leclercq, 2005; Uzan & Lehoucq, 2005).

A quick glance at the contents of different tables of such constants, however, should be sufficient to raise up several questions, bearing upon some deep conceptual aspects of physics. Consider, for instance, the seminal article written in 1929 by Birge, one of the first specialists in the systematic investigation of the physical constants (Birge, 1929). In this paper, which rightly starts by asserting that "some of the most important results of physical science are embodied, directly or indirectly, in the numerical magnitude of various universal constants", the determination is studied of the following constants: velocity of light c, gravitation constant G, relation of litre to cubic centimetre, normal mole volume of ideal gas V_0, relation of international to absolute electrical units, several atomic weights (H, He, N, Ag, I, C, Ca), normal atmosphere, absolute temperature of ice-point, mechanical equivalent of heat J, Faraday constant F, electronic charge e, specific charge of the electron e/m and Planck constant h plus various "additional quantities" (ratio of e.s. to e.m. units, density of water, Rydberg constant, Avogadro's constant N, Boltzmann constant k, etc.). Thirty years later, Cohen, Crowe and DuMond, in their book, *The Fundamental Constants of Physics* (1957), distinguish "classical constants" such as G, V_0, R, J and F from "atomic constants", while recognizing that "the meaning of the term 'atomic constants' has become increasingly inclusive and indefinite". Twelve years later, Taylor, Parker, and Langenberg (1969) chose as "the fundamental physical constants" the following set: c, e, h, N, a.m.u. (atomic mass unit), m_e, M_p, M_n, k, G. More recently, Mohr, Newell, and Taylor (2016) restrain the list of "universal constants" to c, G, h and (curiously) ε_0 and μ_0, while as of 2015, Wikipedia sticks to c, G, h. The heterogeneity and variability of these lists offers a relevant starting point for our reflections.

Here are some of the questions I will attempt to answer in this chapter:

i   why are there "fundamental constants" in physics and not, for instance, in biology or geology?

ii why are there no such constants in the most classical theories of physics, such as classical mechanics?

iii are not the classical constants of thermodynamics and statistical mechanics, R (or k) and J, less fundamental than the "atomic" modern constants, c and h?

iv is there anything common between a simple unit conversion factor, such as the ratio of litre to cubic centimetre, and a universal constant, such as Planck's?

v   why is the velocity of light c considered as a fundamental constant when, according to its very name, it seems to be associated with a particular class of physical phenomena only, namely the propagation of electromagnetic radiation?

vi what is the meaning of taking the value of some of these constants as unity, as though this value was not to be experimentally determined?

vii conversely, how can one let these constants "go to zero" (or to infinity), as if they were not *constants*, in order to define approximate limiting theories such as Newtonian mechanics ($h\rightarrow0$) and Galilean relativity ($c\rightarrow\infty$)?

viii are all the so-called fundamental constants on the same footing, whether they be masses of elementary particles, coupling constants or unit conversion factors, such as M_p, G and the a.m.u. (atomic mass unit) respectively, to stay within the list of "atomic" constants?

ix why do the contents of the tables of "fundamental" physics constants vary with time, as a simple comparison of various such tables reveals (see earlier)?

I will try to show that the answers to these questions and others rely on the understanding of physical science as a historical process. Only by studying the conditions for the appearance, or disappearance, of physical constants can we understand their nature. Only by emphasizing the variations in status of a given constant can we understand its role. Only by contrasting the opposite effects of theoretical and experimental practices upon the fate of such a constant can we analyse its significance. The present investigation thus takes place within a definite vision of physics, and science in general, as a social endeavour. Its ensuing historicity should then be put into light, even at its seemingly most abstract and formal levels. The case of physical constants thus epitomizes this view, since their constant numerical values make sense only through a changing conceptual nature.

1   A classification of physical constants

Let me start by proposing a classification of physical constants into three types. This, hopefully, may bring some order in the otherwise rather incongruous lists offered by the standard tables. By order of increasing generality, I will thus distinguish three types of constants:

A *Physical properties of particular objects:* for instance, the masses of fundamental particles, their magnetic moments, energy widths of unstable ones, etc.

B *Constants characterizing whole classes of physical phenomena:* these are mainly the coupling constants of the various fundamental interactions, such as Newton's constant G associated with gravitation.

C *Universal constants,* such as c or h, which enter the most general theoretical framework available, independently of particular objects or specific interactions (I will come back later on to the validity of approximate theories, where such constants may be neglected – see Section 3).

The interest of such a classification is not to offer an intrinsic, absolute and invariant characterization of any given constant. Quite the contrary, it is its strong time dependence that makes it useful for discussing the changing status of most physical constants. Indeed, mobility is the rule, a constant moving from one type to another when our physical knowledge increases. Consider first the constants of type-A. While for quite a few decades the masses of the nucleons (for instance) belonged to that class, we are now convinced that their values can (or could) be explained in terms of the masses of their constituents (quarks, gluons) and the strengths of their interactions. The nucleon masses, thus, in some sense may be dropped out altogether from the table of fundamental constants, their status now being that of derived quantities. This is the case, even though one is not yet able to compute them exactly from the deeper theory: the principle of their dependence upon more fundamental quantities is sufficient to ensure their "de-fundamentalisation". This is precisely what has happened before to the old physical constants, which, at the beginning of this century, consisted of the macroscopic properties of the simple elements, such as density, compressibility or heat capacity. Now we know that their values rely on the atomic structure of matter and are explainable in principle from quantum theory, even though very few such calculations may be achieved in fact. The same happened for the atomic and molecular properties, such as ionization energies or polarizability, and then for nuclear properties, such as masses, sizes, etc. The new fundamental constants, in terms of which the old ones are explained away, may belong to any of the three classes. The atomic and molecular quantities thus are eliminated once they are known to depend on the electronic mass m (type-A), on the electromagnetic coupling constant e (type-B) and on the Planck quantum constant (type-C). In the same way, the advent of quantum electrodynamics has enabled us to express the electron magnetic moment in terms of the same quantities, so that it is no longer a fundamental constant.

But a type-A constant, instead of vanishing from the table, may be promoted to another category. This is the case of e, for instance. First characterized as the electric charge of the electron, a specific property of a particular object, it was later on recognized as the coupling constant of electromagnetic fields to all charged fundamental constituents of matter and associated with the whole class of all electrodynamic phenomena. It thus became a type-B constant. An even more important example is afforded by the change in status experienced by c. As the terminology unfortunately still reflects, c was first introduced as the speed of light, that is a type-A constant. With the development of electrodynamics (classical), it came to be understood as playing a role in all electromagnetic phenomena: in most theoretical expressions, its significance is not directly that of a velocity (even though its dimensions are, of course), and one might thus think of c as a type-B constant. But the advent of Einsteinian relativity forces us to associate c with the theoretical description of space-time itself, independent of its specific contents. This is proved by the fact that Einsteinian relativity, according to our present knowledge, rules all fundamental interactions, implying the occurrence of c in the relevant theories, even when no electromagnetic phenomena are to be considered at all. This point is blurred by the traditional terminology ("speed of light"), associated with an operational interpretation of relativity theory, whereby the Lorentz transformations are derived from an analysis of communication through electromagnetic signals. The theory, however, may be built upon a structural analysis of spacetime, without using any postulate about the velocity of light (Lee & Kalotas, 1975; Lévy-Leblond, 1976a; Lévy-Leblond & Provost, 1979). That c thus has to be considered as a type-C constant, and not a type-A only, may be further emphasized by stressing that it could well be the velocity of . . . no existing physical object. If the photon had a non-vanishing mass, however small, its velocity would be closely approximated by c in all presently known situations, but would differ from it for low enough energies (Goldhaber & Nieto, 2010). While such an occurrence would not *per se* ruin the validity of Einsteinian relativity, it would, however, invalidate most of its customary derivations. As a last argument, one might think of how Planck constant would have been considered, had it been first introduced through the discovery of angular-momentum quantization for the photon; it would then probably go by the name of the "spin of light". In fact, c is not the speed of light any more than $\hbar$ is the spin of light. Renaming it the "Einstein constant" would certainly be appropriate

The same phenomena of elimination or promotion may affect type-B constants. Indeed, if a theoretical unification of two classes of interactions is realized, one (or perhaps both) of the coupling constants will lose its (their) fundamental nature, in favour of the other one (or of another, new, constant). Such a phenomenon was witnessed in the past with the unification by Maxwell of electricity and magnetism, whereby the magnetic permeability μ_0 and the electric permittivity ε_0 of the vacuum (type-B constants) were found to be related through the "speed of light" c (type-A constant). The same situation in some sense is realized by the unification in the Standard Model of the weak and electromagnetic constants. In the case in which all four (or more) fundamental interactions would be unified, they would be described by a new fundamental universal constant, of type-A.

A similar case would be realized if some of the constants were shown *not* to be constants, exhibiting a cosmological time dependence as in the now abandoned

Dirac hypothesis (Dirac proposed a time variation of the gravitational constant G). Astrophysical observations seem to rule out the idea; see (Uzan, 2011). The new constant parameters in terms of which the time dependence of these no longer constants would be expressed, should then take their place in the tables.

Not only does the type of a fundamental constant (or its absence thereof) depend on the history of physics, but it may also vary according to one's implicit epistemological position. This remark is already clear from our discussion of c. But a better case in point, since a more controversial one, is that of Newton's gravitational constant G. According to the standard point of view upon general relativity, space-time itself is ruled by gravity along with all phenomena within it. General relativity, in its geometrical interpretation, is an all-embracing theory, and its characteristic constant G should thus be elevated to type-C dignity. However, there exists an heterodox point of view, according to which the so-called "general relativity" is but a particular theory of a spin-2 classical field (Weinberg, 1972; Misner, Thorne, & Wheeler, 1973). This field is universally coupled to energy, including its own, which endows it with a specific nonlinear behaviour. Because of this universal coupling, furthermore, the field plays the role of an effective variable space-time Riemannian metric ruling all physical phenomena. Needless to say, the formal theory is exactly identical to the conventional one, so that no experimental discrimination is possible. The advantage of such a view is to maintain gravitation at the level of the other fundamental interactions, its theoretical description being given by a local field theory as well. The price to be paid is the loss of the *a priori* intrinsic geometrical interpretation. Conversely, this unconventional point of view offers more room for modifying the theory if experimental results someday require such a change. In any case, within this framework, G keeps its pre-Einsteinian type-B status, on the same footing as other coupling constants.

Let us, from now on, concentrate on the universal physical constants (type-C).

2   The fate of universal constants

2.1   *Conceptual synthesis and analysis*

In order to understand the role of the universal physical constants, let us consider the particular case of Planck constant h. It was first introduced into physics through the Planck-Einstein relationship, $E = h\nu$. This relationship is customarily interpreted as associating an energy E with the frequency ν of an undulatory physical phenomenon. The connection thus established between a concept of particle mechanics, the energy of a discrete entity and one of wave theory, the frequency, leads to the idea of waveparticle duality in quantum physics, and, further on, to the philosophy of complementarity. Such an interpretation was quite natural in the early days of quantum theory, when one had to approach this new unknown theory from the old classical ones. Duality and complementarity served the very useful purpose of letting physicists use the classical concepts in the quantum domain as far as possible, while taking into account their limits of validity as imposed by these general principles. In such a way, many quantum results were obtained, or at least qualitatively understood, without using a yet-to-be-developed full quantum theory. Most of Bohr's theoretical work is a magnificent example of such a line of thought. It is to be realized today, however, that quantum theory does exist and that its concepts, after a century of collective practice, are deeply rooted in the present common sense of working physicists. These concepts need no longer be approached from classical ones, but may, and should, be taken at their face value. Such an intrinsically quantum understanding leads one to recognize that the objects of quantum physics are not either waves or particles, as duality would want us to believe; they are *neither waves, nor particles*, even though they do exhibit, under very particular circumstances, two types of limit behaviour as (classical) waves, or (classical) particles (see Section 3). It has been proposed to stress this ontological point by calling them "quantons" (Bunge, 1973; Lévy-Leblond, 1976d, 2003). Coming back to Planck's constant, the relationship $E = h\nu$, according to this point of view, is not to be interpreted as linking two classical concepts, but rather as transcending them through their synthesis, to establish a new single concept with a broader scope. The quantum energy indeed is a new concept, since it associates to any physical state a whole spectrum of numerical values and has to be represented by a Hermitian operator, as opposed to the single-number function that represents energy in classical mechanics. Here again a new name should have been given to stress the emergence of this concept, as an intrinsic one. Energy and frequency then appear as two particular facets of a more general notion, each of which being the only visible one from either one of two quite specific points of view.

The role thus played by the Planck constant in bringing together these two facets is characteristic of universal constants. Any universal constant may be so described as a "concept synthesizer", expressing the unification of two previously unconnected physical concepts into a single one of extended validity. It will be shown that classical constants such as k and J play the same role. In any case, the same analysis may be applied to c, one use of which, for instance, is to bring together the concepts of spatial intervals Δx, on the one hand, and time intervals Δt, on the other hand. These are but two aspects of the more general notion of a space-time interval $\Delta s = (\Delta t^2 - c^{-2}\Delta x^2)^{1/2}$, which reduces to one or the other under special circumstances. Any universal constant usually brings about several such synthetic concepts. Planck's constant also unifies momentum and wave number through the de Broglie relationship $p = h/\lambda$, while c unifies mass and energy through the Einstein relationship $E = c^2 m$. This is easily understood, since any physical concept by essence belongs to a theoretical framework which relates it to other concepts. The synthesis of two concepts thus is a local aspect of a more global unification of two pre-existing consistent theoretical structures. Bringing them into contact at one point usually requires the fitting together of other parts as well. This is what happens when the spatiotemporal consistency of particle mechanics, on the one hand, and wave theory, on the other, requires h to play the same role with respect to momentum and wave number (space aspect) as it does with respect to energy and frequency (time aspect). Fully endorsing this point of view may lead to a better understanding of the new concepts. For instance, it enables one to stress that the de Broglie relationship, $p = h/\lambda$, exhibits the intrinsically nonclassical nature of the quantum concept it establishes. Indeed, the classical wavelength is independent of the reference frame, or Galilean invariant, while the momentum of a classical particle with mass m changes according to $p' = p + mv$, under a change of Galilean frame with velocity v. The de Broglie relationship thus seems inconsistent with Galilean invariance, that is, with the structure of spacetime in "nonrelativistic" physics (Landé, 1975). This pseudoparadox, far from dismissing the de Broglie relationship as Landé maintained, points to a conceptual difference between classical wave number which indeed is Galilean invariant, and quantum "wave number" which is not (Lévy-Leblond, 1976c). The quantum "wavelength" transforms according to $1/\lambda' = 1/\lambda + mv/h$, which does *not* reduce to the classical limit $\lambda' = \lambda$, when h "goes to zero" (see Section 3). This fact is related to the quantum "waves" being represented by complex numbers, in contradistinction to the real amplitudes of classical waves (whether they be acoustical, or hydrodynamical). As another example of the accent put on the conceptual nature of h as a building tool of quantum theory, one may write a third relationship on the same footing as the Planck and de Broglie relationships, but concerning now the angular momentum that leads to a heuristic understanding of the discretization of the quantum angular momentum (Lévy-Leblond, 1976b).

Universal constants thus express synthetic transcending not of isolated pairs of concepts, but of whole conceptual arrays. In this sense, a universal constant is a "theory synthesizer", more than a mere "concept synthesize". From this abstract point of view, the various specific syntheses expressed through a universal constant between various pairs of concepts belonging to two theoretical frameworks are but equivalent consequences of the general theoretical unification of these frameworks. Nevertheless, because of historical considerations and epistemological motivations, they are not actually given an equal status, especially in educational practice. Some of them are taken as a starting point or fundamental hypothesis, such as $E = h\nu$, or $\Delta s = (\Delta t^2 - c^{-2}\Delta x^2)^{1/2}$, while other ones are considered as derived relations, or consequences, such as $p = h/\lambda$, or $E = c^2 m$. Since one has to start somewhere, it is probably true that the equivalence of all such expressions, as reflecting various aspects of one and the same synthetic process through a given universal constant, is bound to remain a rather abstract statement. Its acceptance, however, may pave the way to a modification of the traditional hierarchy. As an example, it has been proposed to develop Einsteinian relativity by starting directly from the mass-energy relationship $E = c^2 m$, by building upon it the "relativistic" concepts of energy and momentum, and then by deriving from them the theoretical structure of space-time (Davidon, 1975). After all, this corresponds more closely to the real needs of physics where the Lorentz transformation formulae or invariant expressions are actually more often used for momentum-energy quantities than for space-time ones. Moreover, the role of c in relating energy and inertia (rather than mass) is deeply rooted in the immediate prehistory of Einsteinian relativity, and it "could" have been the starting point of an alternative historical path towards this theory. These considerations, clearly, are of some epistemological and pedagogical importance (Lévy-Leblond, Gazeau et al., 2004).

We are now in a position to answer the question of the existence of universal constants in physics as distinct from other sciences. This is simply because in physics alone do the scientific concepts show an intrinsically mathematical expression. In physics, mathematics does not simply apply; it plays a far deeper, constitutive role (Lévy-Leblond, 1992). The identification, or synthesis of two concepts in physics, thus requires first their mathematical nature to be identical (scalars or vectors, for instance) and then implies the existence of a proportionality factor. Let me stress that numerical measurements of a given quantity, as may exist in other sciences (even social ones), are not sufficient to endow it with mathematical constitutivity; it is necessary that there exist nontrivial mathematical relationships between several such quantities, expressing the "scientific laws" of the field.

But the role of universal constants in the synthesis and unification of previously unrelated concepts or sets thereof, if it is the prime one in historical order of appearance, has for a corollary the fact of their leading to split and separate previously fused, if not confused, concepts. Two simple examples in relativity theory may be given here. The first one deals with the impossibility in Einsteinian relativity of a concept with the following two properties possessed by the velocity in Galilean relativity: (a) being an additive quantity, obeying the simple composition law $v_{12} = v_1 + v_2$, and (b) giving the time rate of spatial change, namely $v = dx/dt$, for uniform motion. In Einsteinian relativity, if the second property is used as a definition of what we will keep calling "velocity" v, the first one will hold true for another quantity, the so-called "rapidity" φ. The two quantities are related by $v = \tanh\varphi$, or, with dimensional notations, by $v = c \tanh(\varphi/c)$, which makes apparent their fusion in the limit $c \to \infty$. The introduction of the concept of rapidity is of a major help for educational purposes (Taylor & Wheeler, 1966). It not only yields a more compact and more significant expression for Lorentz transformations via hyperbolic functions, but it explains away the pseudo-paradoxes associated to the idea of a limiting velocity or nonadditivity of velocities, as simply due to a bad choice of parameter, such as would occur if rotations were labelled through the tangent of the angle instead of the angle itself. In recent decades, the concept of rapidity has also been fruitfully used in high-energy phenomenology (e.g. in Wong, 1994, p. 24). A similar clarification may be achieved in relativistic dynamics, by introducing, with the concepts of energy and mass, the one of inertia, defined as the coefficient of the velocity in the expression for the moment. It is seen then that inertia is to be identified with energy in Einsteinian relativity, but with mass in Galilean relativity. The occurrence of the universal constant c thus splits inertia from mass as it fuses it with energy. In the description of space-time, the splitting of the categories of simultaneous pairs of events from that of null (lightlike) intervals may be interpreted in quite the same way. Other examples can be found at will. To use the same material metaphor as before, it may be said that the fitting of two conceptual structures, while bringing into contact previously separated pieces, also generates stresses requiring various splits within the new body.

2.2   Units and unity

It is an elementary, but crucial, remark that the role of a universal constant as underlying the foundations of new concepts systematically decreases in importance when time goes on and the novelty of the concepts fades away. Indeed, when a sufficient familiarity has been acquired through years of experimenting, theorizing and teaching, one no longer needs to reach these concepts through the relationship of the ancient ones as synthesized by the universal constant. One simply uses the concepts as such. The constant then appears as a mere numerical conversion factor, enabling one to express a given physical quantity in terms of various units. No deep conceptual role is any more attributed to the constant, since the synthesis it symbolizes is, so to speak, achieved from the start. In other words, the theoretical status of a universal constant decreases as its practical importance increases. A good example of this situation is afforded by the classical thermodynamical constants J and k. The first one served to unify heat with work through the relationship $W = JQ$, while the second one showed that temperature was but a statistical aspect of kinetic energy, as expressed by $E = kT$ (up to some numerical factor depending on the number of degrees of freedom). Of course, as emphasized earlier, J and k not only introduced new concepts, but whole new theories: thermodynamics for the first one, statistical mechanics for the second. We are so accustomed today to these ideas that they are incorporated into the implicit background of physical theory. Theoreticians almost automatically choose a convenient system of units such that $k = 1$, since they know that energy and temperature, or work and heat in fact *are* (now) but a single concept. In such a way, these constants gradually fade out of sight in quite a literal sense: less and less are they written in formal expressions. From this point of view, it is seen that J and k indeed are universal constants, in the very same fundamental way as h or c. Only does our long collective practice of the concepts they express enable us to forget about their nature and to consider them as mere conversion factors. It must be said that such a process today is well underway concerning h and c. While all textbooks and articles in the first decades of the twentieth century kept a detailed record of all h's and c's in their formulae, it is the common use today to take them as unity, which only means adopting a more adapted system of units. This convention has become almost tacit in recent years, so that, except perhaps at the educational level, it will soon be obvious that there is no difference of nature between h or c, on the one hand, and k or J, on the other.

This then is the ordinary fate of universal constants: to see their nature as concept synthesizers being progressively incorporated into the implicit common background of physical ideas, then to play a role of mere unit conversion factors and often to be finally forgotten altogether by a suitable redefinition of physical units. Once this is realized, one may well ask how many of these forgotten universal constants are lying around. Let us recall here the "Parable of the Surveyors" due to Taylor and Wheeler (1966):

Once upon a time, there was a Daytime surveyor who measured off the king's lands. He took his directions of north and east from a magnetic compass needle. Eastward directions from the centre of the town square he measured in meters (*x* in meters). Northward directions were sacred and were measured in a different unit, in miles (*y* in miles). His records were complete and accurate and were of often consulted by the Daytimers.

Nighttimers used the services of another surveyor. His north and east directions were based on the North Star. He too measured distances eastward from the centre of the town square in meters (*x'* in meters) and sacred distances north in miles *(y'* in miles). His records were complete and accurate. Every corner of a plot appeared in his book with its two co-ordinates, *x'* and *y'*.

One fall a student of surveying turned up with novel openmindedness. Contrary to all previous tradition, he attended both of the rival schools operated by the two leaders of surveying. At the day school, he learned from one expert his method of recording the location of the gates of the town and the corners of plots of land. At night school, he learned the other method. As the days and nights passed, the student puzzled more and more in an attempt to find some harmonious relationship between the rival ways of recording location. He carefully compared the records of the two surveyors on the locations of the town gates relative to the centre of the town square.

In defiance of tradition, the student took the daring and heretical step to convert northward measurements previously expressed always in miles, into meters by multi plication with a constant conversion factor, K. He then discovered that the quantity $[(x_A)^2 + (Ky_A)^2]^{1/2}$ based on Daytime measurements of the position of gate A had exactly the same numerical value as the quantity $[(x'_A)^2 + (Ky'_A)^2]^{1/2}$ computed from the readings of the Nighttime surveyor for gate A. He tried the same comparison on the readings computed from the recorded positions of gate B, and found agreement here too. The student's excitement grew as he checked his scheme of comparison for all the other town gates and found everywhere agreement. He decided to give his discovery a name. He called the quantity $[(x)^2 + (Ky)^2]^{1/2}$ the *distance* of the point (x, y) from the centre of town. He said that he had discovered the *principle of the invariance of distance;* that one gets exactly the same distances from the Daytime co-ordinates as from the Nighttime co-ordinates, despite the fact that the two sets of surveyors' numbers are quite different.

Now we may realize that there is at least a most important universal constant in the simplest of all physical theories, namely plane Euclidean geometry. This constant expresses the theoretical assertion of space isotropy, enabling us to synthesize the concepts of northward and eastward distances into the single general concept of distance, independent of the orientation. Obviously, the numerical value of the constant K is without physical significance and due to mere historical contingencies. Its very existence, however, is a fundamental aspect of geometry. It is quite clear that, in due time, both distances, eastward and northward, in the town of the parable came to be measured with the same unit, so that the constant K disappeared, and the discovery of the bright student faded into oblivion: was it not an "obvious" result?

This parable was intended by its authors to stress, and rightly so, the analogous role played today by the constant c with respect to spacetime. But the parable of the surveyors also compels us, conversely, to unearth many of these forgotten universal constants, incorporated as they are into what now seems to be immediate truth, but was once the object of a lengthy and difficult working out. Another purely geometrical example is given by the evaluation of areas. Indeed, let us choose a unit of area, as the area of some arbitrary plane figure, for instance an average human hand palm. It is then a "law of physics" (or a theorem of geometry) that the area A of a square with a side of length L, measured with some length unit, for instance the foot, is given by $A = \alpha L^2$, where α is some universal constant (with our units its value is roughly $\alpha = 0.11$ hand palm per square foot). One *then* redefines the unit of area as the area of the square with unit side, so that α vanishes from sight. It should not be forgotten, however, that this constant expressed the now "obvious" synthesis of areas with squares of lengths. The same could be said of volumes, of course, and is not without factual relevance today. After all, the Anglo-Saxon traditional units of volumes, gallons or pints, are *not* defined by cubing the lengths units, foot or inch; the universal constant β entering the relationship $V = \beta L^3$ has not been taken as unity. We will see that even within the scientific metric system, the constant β cannot be forgotten altogether. Now it should be clear that many such hidden universal constants lie at the core of the main statements of classical physics, at most in the oldest theories such as geometry or Newtonian mechanics, in which a long practice has led to the complete incorporation of their significance at an allimplicit level. The absence of universal constants in this part of physics is but an apparent privilege of old age. One might thus classify the universal constants (type-C) into three subclasses according to their historical status:

i   the *modern* ones, such as h and c, the conceptual role of which is still dominant;

ii the *classical* ones, such as k or J, which today appear essentially as unit conversion factors, their conceptual role having become almost implicit; and

iii the *archaic* ones, which have been so well assimilated and digested as to become totally invisible.

2.3 The point of view of practice

The story, however, is not that simple. It is only from the theorist's point of view that the life of a universal constant reaches the happy end of such a drift into the Nirvana of unity and oblivion. The experimentalists working in the laboratory, when making measurements, must use concrete definitions of their units and cannot at will identify two operationally independent standards as the theorists on the paper do. It is a fact that, whatever fundamental system of units is adopted, based on the theoretical knowledge of the time, the use of units belonging to various other previous systems adapted to such and such domain of physics cannot be eliminated together. There are two reasons for this state of affairs. The first one is historical social inertia, which, for instance, forces the experimental physicists on the West side of the Atlantic to plan and order the nuts, bolts, plates, rods, etc. of their apparatus, by stating their dimensions in feet and inches rather than in metres and centimetres. The universal constant x entering the relationship $L_{US} = L_{EU}$ between the length of some object in the United States and the length of the same one in Europe (so that the subscript "EU" refers to some of us, while "US" refers to some of you) thus can be taken as unity in principle – but in principle *only*. This constant, once more, does express a fundamental law of physics, namely the homogeneity of space, enabling one to define the concept of length of an object independently of its location. But there is a second reason for the persisting of nonorthodox units, which is due to the nature of experimental physics proper, of which metrology is a fundamental aspect. Once a system of units is chosen (such as the modern International System of Units), every unit of a physical magnitude not belonging to the fundamental ones may be derived from the fundamental units. However, these derived units are often defined in such a way as to be of a very awkward use, or even as to lack the required precision, in a given domain of physics. Easier and better measurements may be done with the use of an independently defined local system of units. It then remains as a task for experimental metrology to relate these local units to the fundamental system through the experimental determination of conversion factors, which, as shown before, are nothing but genuine universal constants. A simple example may be given here. Before 1964, the litre as a unit of volume was defined independently of the length unit (metre) as the volume of a kilogram of water at 4°C. The relationship between volumetric measurements in litres and linear measurements in metres thus required the experimental determination of the universal constant in the relationship $V = \beta L^3$ between volumes and lengths; the value of this constant was $\beta = 1.000028 \pm 0.000004$ litre·dm^{-3}, which, of course, could be taken as unity by the theorist.[2] It must be pointed out that the "noble" universal constant c is not different in principle from the "trivial" β. At the experimental level considered here, it is to be realized that the possibility of direct measurements of c only comes from it being also a type-A constant; a direct determination of the velocity of photons thus leads to the value of c. But if there did not exist particles with zero mass (or approximately zero in Goldhaber & Nieto, 2010), such a measurement would be impossible and c would have to be indirectly determined by relating measurements of electrostatic and magnetostatic quantities, or of lengths and frequencies. Such indirect determinations of c in the past sometimes have yielded the better values available at the time (Sanders, 1965). The present point in fact has been expressed by the best craftsmen themselves, such as Cohen and DuMond (1965).

Exactly as the practical imperatives of experimental physics prevent a naïve dismissing of "classical" constants by a change in unit conventions, more general social conditions can impose the persistence of "unnatural" archaic constants. Two simple historical examples of such a situation can be offered.

The first one is, once more, the question of volume measurements. Its scientific, metrological aspect, discussed earlier, corresponds to a much more general fact, valid since the highest antiquity; namely volumes usually are *not* determined

by geometrical means from length measurements. Indeed, most volume determinations in practical life concern flowing materials, liquid or granular, such as
beverages or grains (solids are mainly considered according to their weight).
This is why independent volume units, defined by the capacity of some standard
containers, have been the rule until the advent of the metric system (and there-
after). For instance, in the Anglo-Saxon system, it has been said, pints are not
related to inches. As a consequence, even though, as far as order of magnitudes
are concerned, the volume unit usually has been comparable to the cube of the
length unit, there was no real need to redefine these units in such a way that the
appropriate universal constant be unity.[3] Surface measurements offer a comparable
instance, with special units for land areas, although it is much less marked than for
volumes since most determinations of surfaces have been done through length
measurements by geometrical means in all known historical periods.

The second example is furnished by the constant expressing the homogeneity of
space, that is, the possibility of using the same units of length at each and every
point of space. Apart from the case of the Anglo-Saxon system, this constant is
now almost universally taken as unity, due to the international adoption of the
metric system. This is a recent occurrence on the historical scale, however, and
beyond a doubt came about very much later than the theoretical understanding of that possibility. The point is that such a unification was not necessary
since space indeed was heterogeneous, socially, if not geometrically. Localized
and rather autonomous social entities, from tribes to cities, were the rule in the
human society until less than a few centuries ago. The progressive unification of
the social space is related to the rise of merchant and industrial capitalism. It is
the need of this new social order which brought about the redefinition of local
units of lengths, so that the corresponding universal constant took on the status
of an archaic one.[4]

3 The case of the vanishing constants

3.1 *How to vary a constant*

Universal constants not only play a role as standards of definition and measurement for physical quantities. They are further used as standards of validity
for physical theories. This aspect is usually summarized by statements such as
"Galilean relativity is obtained from Einsteinian relativity in the limit $c \to \infty$",
or "quantum mechanics goes into classical mechanics when Planck constant vanishes". Now these clearly are rather loose assertions, which are of formal significance at most, as they bear upon purely mathematical limiting processes imposed
to the equations of the theory. But in the real world, the universal constants take
on definite values and one is not free to change them at will. A better way of
expressing these ideas is to assert the validity of Galilean relativity (respectively,
classical mechanics), whenever c (respectively, h) can be considered as very large
(respectively, small). One, however, has to be more accurate: large (or small)
with respect to what? In the case of relativity, it is usually stated that the Galilean

theory holds good, whenever the velocities are small with respect to c. But this is a necessary condition only, and it may be shown that it is not sufficient. Lorentz transformations with small velocities compared to c are approximated by Galilean ones, only for spatiotemporal intervals (or, more generally, four-vectors), which are of the "large timelike" type, that is, such that $\Delta x \ll c\Delta t$. In the opposite case, that is for intervals (or four-vectors) of "large spacelike" type, that is, such that $\Delta x \gg c\Delta t$, an alternative limiting behaviour is obtained, giving rise to a second "nonrelativistic" limit of the Poincaré group, the Carroll group (Lévy-Leblond, 1965). True, the space-time intervals concerned by such transformations, as, for instance, the interval between your reading of the next comma here and now, and the emission within a second of a photon from some far-away star at a distance of 2000 lightyears from home (Jagger & Richard, 1966), are between events with no possible causal connection, precisely because of the large spacelike nature of these intervals. The Carroll group, thus, necessarily applies to an acausal world (hence its name) and its physical relevance is dubious, to say the least (unless tachyons exist, the "nonrelativistic" properties of which could then be described through a Carrollian theory).

However, the very existence of the Carroll group, a well-defined and consistent mathematical, if not, physical object, serves to point out the necessity of a more stringent statement about the condition of validity for the Galilean approximation in relativity theory. It is to be required in effect that *all* relevant physical quantities with the dimensions of velocity (LT^{-1}) be small compared to c, that is, not only actual velocities of moving objects, but ratios of spatial to temporal intervals, of energies to momenta, etc. The necessity of such a general explicit condition may have been blurred by the narrow interpretation of c as a mere velocity (see earlier). Once it is recognized as a truly universal constant, it clearly acts as a standard of comparison for all physical quantities with the same dimensionality. At least this requirement is imperative if one is to set up a consistent theory. Weaker requirements may be sufficient to deal with specific situations. As an example, consider the transformation properties of electromagnetic fields under Lorentz transformations. If the velocity is low enough compared to c, the following formulae hold good:

$$\mathbf{E}' = \mathbf{E} - \mathbf{v} \times \mathbf{B}, \ \mathbf{B}' = \mathbf{B} - c^{-2}\mathbf{v} \times \mathbf{E} \tag{1}$$

Now these formulae cannot fit into a full theory of electromagnetism in agreement with Galilean relativity (Le Bellac & Lévy-Leblond, 1973; Rousseaux, 2013). In such a theory, two types of electromagnetic fields may exist with respective transformation properties:

$$\mathbf{E}' = \mathbf{E} - \mathbf{v} \times \mathbf{B}, \ \mathbf{B}' = \mathbf{B} \tag{2}$$

or

$$\mathbf{E}' = \mathbf{E}, \ \mathbf{B}' = \mathbf{B} - c^{-2}\mathbf{v} \times \mathbf{E} \tag{3}$$

It is clear that (2) or (3) is valid according as $E/B \gg c$ or $E/B \ll c$ (in addition to $v \ll c$). These remarks, obviously, are related to the idea stressed earlier that a universal constant does not underlie a single concept, but a whole theoretical framework.

The situation has been clearer in that respect for quantum mechanics. Since h was never confused with a type-A constant (the "spin of light"), from its universal nature it was rightly inferred that it had to be small compared with all relevant physical quantities with the dimension of an "action" (dimensionality ML^2T^{-1}) for classical mechanics to be approximately valid. There is, however, a number of unsolved problems about the relationship of quantum theory to its classical limit(s), as will be mentioned later.

Another way of expressing the smallness (or largeness) of some universal constant in a given physical situation is to consider the units appropriate to the description of that situation; that is, if they exist at all, units such that all of the physical quantities take on "reasonable" values, spanning a limited range around unity. If the universal constant, when expressed with these units, is very small (or very large), then the approximate theory is valid, which corresponds to the limit in which the constant goes to zero (or to infinity). This is clearly the case in the two examples mentioned up to now, where h takes on a very small value and c a very large one, when expressed in any system of units adapted to our daily experience (whether it be SI, or CGS, or the traditional Anglo-Saxon non metric system).

The last remark, trivial as it may seem when applied to our modern familiar and revered universal constants h and c, may be of some help in under standing the historical reasons for the emergence, and later subsidence, of most universal constants, including the classical and archaic ones. Indeed, for c to appear as a universal constant, it was necessary for experimental investigation to come to grip with some phenomenon where at least one combination of physical quantities with dimension LT^{-1} was comparable to c. This required a stage in the development of experimental techniques which was not reached until the seventeenth century with the first measurements of the velocity of light (Sanders, 1965). Spatiotemporal ratios were for quite a time the only magnitudes with the required dimensionality to be measured with the necessary precision, so that c could not appear but as a type-A constant: the velocity of light and nothing more. It was not well until the nineteenth century that other physical magnitudes, namely electromagnetic ones, could be measured with a sufficient precision. Magnetism, after electricity, was subjected to accurate definitions and measurements, and the remark was left to Kirchhoff (1857) and Riemann (1867) that the combination of electric and magnetic constants, which in modern formulation we would write as $(\varepsilon_0\mu_0)^{-1}$, was quite close to the speed of light. This first hint that c could well be at least a type-B constant, characteristic of electrodynamics in general, was confirmed by Maxwell's achievement of a consistent theory. By the beginning of this century, experimental progress had been such as to yield a vast number of combinations with the dimension LT^{-1}, the values of which were no longer small with respect to c; not only ratios of space to time intervals, but also of electric to magnetic field strengths (starting with Hertz's experiments on electromagnetic radiation), products of frequencies and wavelengths, ratios (square root of) of energies to masses (in the energetics of nuclear reactions), of energies to momenta (dynamics of charged particles), etc. As for h, its existence could not be inferred before the possibility of investigating phenomena where characteristic "actions" were small enough and could be determined with a sufficient precision. The first of these turned out in spectroscopic studies when the black body spectrum was studied at temperatures such that the maximum in the energy of the emitted radiation fell into an accessible range of wavelengths; namely, for Wien's laws to be discovered, the combination $kT\nu^{-1} = kT\lambda c^{-1}$ of the physical parameters under control had to be small enough to become comparable with h. Then the photoelectric effect disclosed as well the presence of the Planck constant when the emission of ultraviolet radiation was mastered: the ratio of the kinetic energy of liberated electrons to the frequency of the radiation, that is $E\nu^{-1}$, could be measured with an accuracy enabling it to be compared with h (Lévy-Leblond, 2000) as classical universal constants are concerned; J could not appear before a theoretical definition, and experimental measurement of heat had to be pushed up to the point at which heat quantities ΔQ could be compared to the amounts of work ΔW commonly occurring (Carwell & Hills, 1976; Pacey, 1974). The development of heat engines, such as the steam machine first, the progresses of physiology as well and, simultaneously, the improvement of thermometry and calorimetry, then led to the recognition of the "equivalence" between heat and work – one should better say their synthesis into a broader concept of energy – by Joule, Mayer, Helmholtz and others (Elkana, 1974).

The hidden character of what was called above the archaic universal constants is readily understood from the present point of view. From the very physiological characteristics and social practice of humanity it follows that isotropy of space; for instance, it has probably been incorporated right from the beginning in the use of one and the same unit of length for measuring distances in all directions. By a blending of Taylor and Wheeler's earlier "Parable of the Surveyors" (Taylor & Wheeler, 1966) with Abbott's well-known novel *Flatland* (1952), one could imagine, however, a science-fiction story of an almost-flat species of intelligent beings. From their direct experience of the world, they could apprehend it on a scale of, say, metres in horizontal planes, but micrometres only in vertical directions. They would certainly use respective units with such a 10^6 ratio for their daily needs in vertical- and horizontal-length measurements. The equivalence of vertical displacements Δz and horizontal ones Δl through a very small universal constant $\eta = 10^{-6}$ μm^{-1} would only be discovered by building telescopes giving access to vertical distances much larger (in the ratio $10^6 = \eta^{-1}$) than the ordinary heights in this world; alternatively, it could be brought to light by the investigation of microscopic horizontal displacements. Conversely, it is clear that two-dimensional physical theories, whether they be purely conceptual exercises or approximate descriptions of some physical phenomena, are in the same relationship with full three-dimensional theories, as, for instance, Galilean relativity to Einsteinian relativity, or classical physics to quantum physics. For such a two-dimensional theory, say, to be valid, all quantities with the dimensions of the ratio between a "vertical" and a "horizontal" length must be small compared with the universal constant η. In our conventional systems of units, of course, η is equal to unity, and we recover a more customary statement for the validity of such approximate theories.

3.2 The limitations of limits

Let us now come back to our conventional modem universal constants in order to examine more closely the significance of limits such as $h \to 0$, or $c \to \infty$. We have already stressed that the constants in fact are constant and that, physically, the limits obtain when the dimensional ratios of the relevant physical quantities, say A/B, are small (or large) compared with the universal constant K which relates (synthetizes) the two quantities A and B. It remains true that, in the formal expressions of the theory, this corresponds to considering the dimensionless ratio KB/A as large (or small), a situation which may be obtained as well giving a large (or small) value to the "constant" K. But several comments must be made to emphasize the limitations of these limit processes, which must be handled and interpreted with some care, if one is to avoid, or to correct, misunderstandings and delusions; the uniqueness, singularity and validity of such limiting processes will be investigated in turn.

Uniqueness. Contrarily to a naïve idea, a given theory does not necessarily possess a unique more restricted theory as a limit, as when it is said that Newtonian mechanics is the "nonrelativistic" limit of Einsteinian mechanics. This is most clearly seen when the theory is expressed in its natural units, in which the universal constant is taken as unity to fully express the concepts of which it underlies the synthetic nature. For then, there is no longer any apparent dimensional constant that might go to zero, or infinity. One has to deal directly with the relevant ratios of physical quantities (dimensionless here), the choice of which must be guided by physical considerations. In other words, there are several non-equivalent ways to reintroduce a constant in the theory, and several corresponding theories when the constant is eliminated through a limit process. The different possibilities deal with differing physical situations. A simple example of this point is furnished by Einsteinian relativity, as already mentioned. Let us write the Lorentz transformation formulae for space-time intervals (Δx, Δt) in units such that $c = 1$

$$\Delta x' = \gamma (\Delta x - v\Delta t) \tag{4a}$$
$$\Delta t' = \gamma (\Delta t - v\Delta x) \tag{4b}$$

where, as usual, we have defined $\gamma = (1 - v^2)^{-1/2}$, v being the velocity of the transformation. It is immediately apparent that the condition $v \ll 1$ is *not* sufficient to yield the Galilean transformations

$$\Delta x' = \Delta x - v\Delta t \tag{5a}$$
$$\Delta t' = \Delta t \tag{5b}$$

One must require in addition that $\Delta x \ll \mathrm{D}t$, that is that the intervals be of the large timelike type. This second condition is necessary for the second term in the equation (4b) to be neglected. If it does not hold, one may only write

$$\Delta x' = \Delta x - v\Delta t \tag{6a}$$
$$\Delta t' = \Delta t - v\Delta x \tag{6b}$$

These, as already said, may be useful approximate formulae in some circumstances, but cannot be the basis of a consistent relativity theory, since they obey no group law and hence no principle of relativity. Now the obvious symmetry of (4a) and (4b) of (4a) suggests a second limit, when $v \ll 1$ and $\Delta x \ll \Delta t$. These low velocity transformations of large spacelike intervals read

$$\Delta x' = \Delta x \tag{7a}$$
$$\Delta t' = \Delta t - v\Delta x \tag{7b}$$

They do obey a group law, defining the so-called Carroll group, already alluded to. The Einsteinian relativity thus possesses two quite different "nonrelativistic" limits, the Galilean and the Carrollian ones. That the second one cannot derive from the usual limit $c \to \infty$ is due to the fact that the conventional dimensional expression of the Lorentz transformation is such as to exclude a Carrollian situation right from the beginning. Indeed the replacements $v \to v/c$ and $\Delta t \to c\Delta t$, which enable one to recover the usual expression from (4) are such that $\Delta x/\Delta t \to \Delta x/c\Delta t$, which necessarily goes to zero (large timelike type) along with $v \to v/c$, when $c \to \infty$, leading to the Galilean transformations. If one was to "dimensionalize" the Lorentz transformations (4) through the replacements $v \to v/c'$, but $\Delta x \to c'\Delta x$, now the limit $c' \to \infty$ would yield the Carrollian transformations (7). The point clearly is that the conventional limit $c \to \infty$ is of a rather tautological nature, since it corresponds to following the evolutionary track of relativity theory in a time-reversed order. It is no surprise that it brings one back to the point of departure: that is the Galilean theory. If one is to study the possible limits of a theory, one must start from this theory as such, expressed within its autonomous system of concepts and intrinsic units. Once more it is seen how much the universal constants, even in the very most technical formulae, bear the mark of the historical developments of physics. The Carroll group probably is of little physical interest, so that the earlier considerations might seem of academic significance. A study of the nonrelativistic (Galilean) approximations to Maxwell's equations, however, leads in much the same technical way to realize that there exists two relevant physical limits (Le Bellac & Lévy-Leblond, 1973; Rousseaux, 2013). There are two Galilean electromagnetisms, depending on whether it is the ratio of electric to magnetic fields, E/B, that is supposed to be small (in dimensionless units), or its inverse. Not surprisingly, one of these limits deals essentially with electric effects, the other one with magnetic effects. It must be mentioned, however, that both go beyond conventional electrostatics and magnetostatics in that they include, for instance, induction phenomena. It is to be said also that possible, more complicated and more interesting Galilean theories of electromagnetism exist that are *not* limits of Maxwell theory, in very much the same way that there exist "nonrelativistic" relativity groups that are not limits of the Lorentz group, such as the Newton groups (Bacry & Lévy-Leblond, 1968). A very elementary example of the existence of two different limits for a given theory is given by ordinary three-dimensional geometry. In writing the spatial interval $(\Delta r)^2 = (\Delta x)^2 + (\Delta y)^2 + (\Delta z)^2$, a universal constant expressing the commensurability of horizontal and vertical lengths may be re-introduced in either of two ways. One may rescale the vertical lengths according to the replacement $\Delta z \rightarrow H\Delta z$, or the horizontal ones by $\Delta x \rightarrow H'\Delta x$, $\Delta y \rightarrow H'\Delta y$. The limits $H \rightarrow 0$ and $H' \rightarrow 0$, respectively, give rise to a two-dimensional plane geometry (see our previous parable) or to a one-dimensional linear one. I will comment on the case of quantum mechanics and the limit $h \rightarrow 0$, since the situation is much less clear. However, it is well known in advance that quantum theory at least has *two* classical limits, dealing respectively with waves and corpuscles.

Singularity. Kuhn (1962) has argued that the history of science proceeds through "scientific revolutions", in between which scientific activity would consist of "normal science". These revolutions would bring about the replacement of old paradigms by new ones, such that the ideas and concepts would undergo radical changes. For instance, according to an admittedly too schematic but common interpretation of Kuhn's ideas, mechanics is supposed to be so affected by the Einsteinian revolution that our ideas on space-time kinematics and dynamics have nothing in common any more with those of Newtonian physics. Such strong statements, obviously, are contrary to all our experience as inquiring and teaching physicists. The difficulty here is that of the apparent dilemma between a continuous view of the history of science, which would deny any qualitative change, and a discontinuous one that ultimately fails to interpret the process of change from one stage to the other. This is not the place to attempt a global evaluation of Kuhn's sociological history of science. One restricted aspect of his views, however, is closely related to the present investigation, namely the nature of the relationship between two successive paradigms in a given scientific domain. Taking as an example Kuhn's case of Einstein *vs.* Newtonian mechanics, let us try to put it into perspective. This historical perspective, it must he emphasized first, needs a backwards look. Obviously, the relationship between some physical theory and a more general successor cannot be studied until the generalization has succeeded. It is then necessarily from the point of view of the new, more encompassing, paradigm that the old one is to be judged. There is no vantage point, outer to both, from which their borderline could be seen and the transition analysed. We have to assess Newtonian mechanics starting from the Einsteinian one. In other words, the epistemological approach is necessarily opposite to the chronological one. It may be suggested then that this approach is that of a *singular* limit, in the mathematical sense of the word, that is a situation where the continuity of a process breaks down for a certain value of the parameter ruling it. This statement, first, is certainly true at the factual level. Indeed the restricted theories, Galilean relativity or, more generally, Newtonian mechanics are obtained from the modern more general "relativistic" theory by a limit process which is necessarily singular. If it were not, the change would amount to a simple rescaling, without any conceptual modification. It is only in the limit in which c goes to infinity, and not when it is arbitrarily large but finite, that the old theory is recovered. In the case of relativity theory, a definite mathematical framework exists, the theory of contraction of groups (Inönu & Wigner, 1953; Saletan, 1961), showing how a continuous family of group isomorphisms depending upon some parameter may tend towards a singular limit, whereby a new, nonisomorphic, group is obtained. But the idea of the old paradigm as a singular limit of the new one is proposed here in a wider, metaphorical sense, as well. It may help understanding how the transition from one to the other, as expressed by the vanishing (or infinity) of some universal constants, brings about qualitative changes into the conceptual tools of the trade. Indeed, if a universal constant brings about the synthesis and the unification of two previously unconnected concepts, its vanishing must be shown to give rise to the converse disjunction, clearly a very singular phenomenon. This is the only way to understand, for instance, how the quantum energy-pulsation branches off into classical particle energy and classical wave pulsation.

As another example, Einstein mechanics knows of only two conserved quantities, energy and momentum, while Newtonian mechanics imposes the further conservation of mass, but introduces another, nonconserved quantity, namely internal energy. In this case, clearly, it is the Einsteinian mass, which in the Galilean limit yields both a conserved mass m and nonconserved internal energy U; of course Einstein's role was precisely to operate the inverse synthesis through the relationship $U = mc^2$! Let it be clear, however, that the singularity may be that of a coalescence of concepts as well as of a disjunction, since we deal here with the converse processes to both the syntheses and the splitting described earlier. But this aspect is rather trivial, consisting for example in the merging of the rapidity φ with the velocity $v = \tanh \varphi$ in the Galilean limit. A final remark, at a more restricted technical level, derives from the mathematical singularity of these limit processes, as considered through universal constants. Much care must be exercised in investigating the limit of some theoretical expression when a universal constant is washed out by letting it go to zero or to infinity. In particular, the units used to write this expression should not depend on the constant itself. Obvious as it may seem, this rule is violated, for instance, by the numerous statements to the effect that "the magnetic moment of the electron, namely $\mu = eh/2mc$, is due to a relativistic phenomenon, because it is seen to depend on c". But this expression for μ is valid only in a system of units where the units of electric and magnetic field strengths are identical, which, as emphasized earlier, cannot be consistent with a Galilean theory. With different units, such as the SI ones for instance, the magnetic moment reads $\mu = eh/2m$ and, being independent of c, should be indifferent to the divergences of opinion between Galileo and Einstein. Indeed, it may be shown that the correct value of the moment obtains as well in a minimal Galilean theory of quantum particles with spin ½ as in Dirac theory (Lévy-Leblond, 1967) (of course, this is because spin itself is still much less an Einsteinian concept). This result extends to higher values of the spin (Hagen, 1970; Hagen &

Hurley, 1970). Simple dimensional considerations show, on the other hand, that higher multipole electromagnetic moments, such as those that exist for high-spin particles, do vanish in the Galilean limit in which $c \rightarrow \infty$ (provided one consider elementary, structureless, particles, and not, for instance, composite systems such as the deuteron, of which the quadrupole moment owes nothing to Einstein, of course). Similar considerations may help in understanding the nature of the spin-orbit coupling for atomic electrons. It is usually said that the Thomas precession factor, due to an Einsteinian effect, halves the conventional coupling of the spin to the apparent magnetic field generated in the electron frame by the Coulomb field of the nucleus. It is very difficult, however, to accept that such a definite factor of ½ might go to 1 in the limit $c \rightarrow \infty$! A more rigorous analysis, in fact, shows that both terms are due to Einsteinian relativity (or else, that in a more complex Galilean theory, they could both exist but be numerically independent). In other words, before assessing the nature of a given effect or a property as due to the specificity of some theory because of its disappearance in the limit theory obtained when the relevant universal constant vanishes, a careful dimensional analysis of the problem is necessary, which requires explicitly disentangling some commonly used conventions of units, when they precisely rely on the theory the limits of which one is to test.

Let us end this section by an unfortunately half-baked idea that might be doomed to failure as an actual programme, but should at least serve to underlie the highly singular nature of the limit processes on universal constants. The point is that, as already mentioned, we usually consider these limits in full knowledge of the resulting theory we want to obtain. The surprise, then, is meagre. Some exceptions have been mentioned (Carroll group, Galilean electromagnetism). But consider a more complicated process in which two universal constants simultaneously are pushed out of the theory. Specifically, let the dimensionless "fine structure" constant of electrodynamics $\alpha = e^2/\hbar c$ keep its value, while $c \rightarrow \infty$ and $\hbar \rightarrow 0$ simultaneously. The result, if there is one, should be a Galilean classical electrodynamics in which only all the quantities of Einsteinian quantum electrodynamics depending on α would keep their values, such as the ratio of the electromagnetic moment to its bare value and all other results of the renormalization programme. True, this bare value, that is $\mu = e\hbar/2m$, goes to zero with $\hbar$, but why not to try computing directly the ratio of the dressed to the bare value, or even rescale m as well? It is very clear that such a theory, challenging as it is, requires a most careful analysis and re-writing of quantum electrodynamics; its singularity is certainly such as to brave any brute-force investigation.

Validity. The last point to emphasize is that the existence of a well-defined formal limit for a theory when some universal constant vanishes is in no way sufficient to guarantee the physical relevance of this limiting theory. The Carrollian kinematics (Lévy-Leblond, 1965) offers an elementary illustration, since it has probably no applicability whatsoever in physical situations. But quantum mechanics offers a much richer and deeper example (Lévy-Leblond, 1976d). Indeed it is a surprise to realize, after almost a century of quantum theory, how little is known on its classical limits. Even at the formal level, things are far from being clear. It is empirically and historically known that quantum theory has resulted from the transcending synthesis of classical wave theory and particle mechanics. One should then be able to recover both these theories as limits of quantum theory now taken as such. The classical particle mechanics limit has received some attention, and various illustrations, from the Ehrenfest theorem to the JWKB approximation, or the relationship with Hamiltonian formalism, may be given of the transition. Things are much less clear on the other side, concerning classical wave theory. Indeed, since the vanishing of the *same* universal constant h seems implied in both limits, some additional assumption has to be made. From the empirical point of view, it is to be realized that an approximate classical particle behaviour may be exhibited by all quantum particles under specific circumstances: bubble or spark chambers thus exhibit "trajectories" and "collisions" for electrons as well as for photons and stranger particles yet. In contradistinction, approximate classical wave behaviour is shown but by boson assemblies, the one important example here being that of the electromagnetic field. It becomes clear then that the classical wave limit requires considering an indefinitely increasing number of particles while the Planck constant vanishes. This point has being given too little attention (Hepp, 1974).

Finally, it is to be emphasized that the existence of such formal limits is by no means a guarantee of the applicability of the approximate theory, or theories, thus obtained. Much more detailed assumptions have to be made if one is to understand and control at the theoretical level the approximate validity of a given limiting theory, even though it may be tested empirically. Concerning the case of quantum mechanics, for instance, we know today that macroscopicity is *not* a sufficient condition for classicality, as is demonstrated by the existence of macroscopic quantum effects (in superfluids, for instance). It is not a surprise then that the very existence of ordinary, hard and stable matter, as approximately described by the classical mechanics of solid bodies, requires a very deep analysis at the quantum level in order to be understood from first principles (Dyson & Lenard, 1967, 1968; Lieb & Seiringer, 2009). It may serve as a useful conclusion by reminding us that understanding the role of the physical constants is but the beginning of a concrete physical analysis, and it only helps in asking the *right* questions – which are now *left* to be answered.

Notes

1 This chapter is a revised and updated version of Lévy-Leblond (1977).
2 Nowadays, the litre has been redefined so as to be identical with the cubic decimetre (see www.bipm.org/jsp/fr/ViewCGPMResolution.jsp?CGPM=12&RES=6).
3 An account of this situation has been given by Casimir (1968) in a humorous parable illustrating the problem concerning electrical units inside dielectric media.
4 Let us not forget, though, that there are some domains where, for good reasons, horizontal and vertical distances are not measured with the same unit, such as air travel, where height is expressed in feet and lengths in miles. Better not to confuse them! More generally, quite a number of recent technological accidents have

been due to errors in unit conversions. Thus, in September 1999, the NASA Mars Climate Orbiter craft was lost due to a ground based computer software that produced output in non-SI units of pound-seconds (lbf s) instead of the metric units of newton-seconds (N s) used by the craft. See http://spacemath.gsfc.nasa.gov/weekly/6Page53.pdf. For another well-known air accident due to a conversion error, see https://en.wikipedia.org/wiki/Gimli_Glider.

References

Abbott, E. A. (1952). *Flatland: A romance of many dimensions.* New York, NY: Dover Publication.

Bacry, H., & Lévy-Leblond, J.-M. (1968). Possible kinematics. *Journal of Mathematical Physics, 9,* 1605–1614.

Birge, R. I. (1929). Probables values of the general physical constants. *Reviews of Modern Physics, 1,* 1–73.

Bunge, M. (1973). *Philosophy of physics.* Dordrecht: Reidel.

Cardwell, D. S. L., & Hills, R. L. (1976). Thermodynamics and practical engineering in the nineteenth century. *History of Technology, 1,* 5–6.

Casimir, H. B. G. (1968). On electromagnetic units. *Helvetica Physica Acta, 41,* 741–742.

Cohen, E. R., Crowe, K. M., & DuMond, J. W. M. (1957). *The fundamental constants of physics.* New York, NY: Interscience.

Cohen, E. R., & DuMond, J. W. (1965). Our knowledge of the fundamental constants of physics and chemistry. *Reviews of Modern Physics, 37,* 537–594.

Davidon, A. (1975). Consequences of the inertial equivalence of energy. *Foundations of Physics, 5,* 525–542.

Duff, M. J., Okun, L. B., & Veneziano, G. (2002). Trialogue on the number of fundamental constants. *Journal of High Energy Physics,* 03 (2002), 023, and arXiv:physics/0110060v3.

Dyson, F. J., & Lenard, A. (1967–1968). A new approach to the stability of matter problem. *Journal of Mathematical Physics, 8,* 423–434; *9,* 698–711.

Elkana, Y. (1974). *The discovery of the conservation of energy.* London: Hutchison.

Goldhaber, A. S., & Nieto, M. M. (2010). Photon and graviton mass limits. *Reviews of Modern Physics, 82,* 939–979.

Hagen, C. R. (1970). The Bargmann-Wigner method in Galilean relativity. *Communications in Mathematical Physics, 18,* 97–108.

Hagen, C. R., & Hurley, W. J. (1970). Magnetic moment of a particle with arbitrary spin. *Physical Review Letters, 24,* 1381–1384.

Hepp, K. (1974). The classical limit for quantum mechanical correlation functions. *Communications in Mathematical Physics, 35,* 265–277.

Inönu, E., & Wigner, E. P. (1953). On the contraction of groups and their representations. *Proceedings of the National Academy of Sciences, 39,* 510–524.

Jagger, M., & Richard, K. (1966). *Two thousand lightyears from home* (Mirage Music, 1966). Recorded by The Rolling Stones in *Their Satanic Majesties Bequest,* TXS 103 (Decca, 1966).

Kirchhoff, G. (1857). Über die Bewegung der Elektricität in Leitern. *Annual Review of Physical Chemistry, 102,* 529–544.

Kuhn, T. (1962). *The structure of scientific revolutions.* Chicago, IL: University of Chicago Press.

Landé, A. (1975). Quantum facts and fiction. *American Journal of Physics, 43,* 701–704.

Le Bellac, M., & Lévy-Leblond, J.-M. (1973). Galilean electromagnetism. *Nuovo Cimento, 14B*, 217–234.

Lee, A. R., & Kalotas, T. I. (1975). Lorentz transformations from the first postulate. *American Journal of Physics, 43*, 434–437.

Lévy-Leblond, J.-M. (1965). Une nouvelle limite non-relativiste du groupe de Poincaré. *Annales de l'Institut Henri Poincaré, 3A*, 1–12.

Lévy-Leblond, J.-M. (1967). Nonrelativistic particles and wave equations. *Communications in Mathematical Physics, 6*, 286–311.

LévyLeblond, J.-M. (1976a). One more derivation of the Lorentz transformation. *American Journal of Physics, 44*, 271–277.

Lévy-Leblond, J.-M. (1976b). Quantum heuristics and angular momentum. *American Journal of Physics, 44*, 719–722.

Lévy-Leblond, J.-M. (1976c). Quantum fact and fiction: Clarifying Landé's pseudo-paradox. *American Journal of Physics, 44*, 1130–1132.

Lévy-Leblond, J.-M. (1976d). On the nature of quantons. *Dialectica, 30*, 161–192.

Lévy-Leblond, J.-M. (1977). On the conceptual nature of physical constants. *La Rivista del Nuovo Cimento, 7*, 187–213.

Lévy-Leblond, J.-M. (1992). Why does physics need mathematics? In E. Ullmann-Margalit (Ed.), *The scientific enterprise.* Dordrecht: Kluwer.

Lévy-Leblond, J.-M. (2000). The meanings of Planck's constant. In E. Beltrametti et al. eds, One Hundred Years of h, *Proceedings of Pavia Conference* (Italian Physical Society).

Lévy-Leblond, J.-M. (2003). On the nature of quantons. *Science and Education, 12*, 495–502.

Lévy-Leblond, J.-M. (2004). What if Einstein had not been there? In J.-P. Gazeau et al. (Eds.), *Group 24: Physical and mathematical aspects of symmetry*, IOP Conf. Series 173 (pp. 173–182). Bristol: IOP.

LévyLeblond, J.-M., & Provost, J.-P. (1979). Additivity, rapidity, relativity. *American Journal of Physics, 47*, 1045–1049.

Lieb, E. H., & Seiringer, R. (2009). *The stability of matter in quantum mechanics.* Cambridge: Cambridge University Press.

Misner, C. W., Thorne, K. S., & Wheeler, J. A. (1973). *Gravitation.* New York, NY: Freeman.

Mohr, P. J., Newell, D. B., & Taylor, B. N. (2016). CODATA recommended values of the fundamental physical constants: 2014. *Reviews of Modern Physics, 88*, 035009, 73 pages.

Pacey, A. J. (1974). Some early heat engine concepts and the conservation of heat. *British Journal for the History of Science, 7*, 135–145.

Riemann, G. F. B. (1867). Ein Beitrag zur Elektrodynamik. *Annual Review of Physical Chemistry, 131*, 237–242.

Rousseaux, G. (2013). Forty years of Galilean Electromagnetism (1973–2013). *European Physical Journal – Plus, 128*(81), 1–14.

Saletan, E. (1961). Contraction of Lie groups. *Journal of Mathematical Physics, 2*, 1–2.

Sanders, J. H. (1965). *The velocity of light.* London: Pergamon.

Taylor, B. N., Parker, W. H., & Langenberg, D. N. (1969). Determination of e/h, using macroscopic quantum phase coherence in superconductors: Implications for quantum electrodynamics and the fundamental physical constants. *Reviews of Modern Physics, 41*, 375–496.

Taylor, E. F., & Wheeler, J. A. (1966). *Spacetime physics.* San Francisco, CA: Freeman.

Uzan, J.-P. (2011). Varying constants, gravitation and cosmology. *Living Reviews in Relativity, 14*(1), 2.

Uzan, J. P., & Leclercq, B. (2005). *De l'importance d'être une constante.* Paris: Dunod.

Uzan, J. P., & Lehoucq, R. (2005). *Les constantes fondamentales.* Paris: Belin.

Weinberg, S. (1972). *Gravitation and cosmology.* New York, NY: Wiley & Sons.

Wong, C.-Y. (1994). *Introduction to high-energy heavy-ion collisions.* Singapore: World Scientific.

7 And how experiments begin

The International Prototype Kilogram and the Planck constant

Sally Riordan

1 The conventionalist account of scientific standards

A straightforward view of scientific standards goes as follows. Equipped with an understanding of what is being measured (length), it becomes necessary, if experiments are to be conducted at all, to "point" to a certain quantity of that substance (one metre). The pointer, or definition of the standard, is given by a stipulation (the distance between the two lines upon this metal bar, when at the temperature of melting ice, shall be one metre). At least one procedure, whether tacitly understood or developed specifically for the occasion, must be available in order to then realize this pointer (the lining a metal bar of approximately one metre against the prototype metre and taking a reading of its length from a measuring microscope). A particular realization may be improved to increase its accuracy (giving a value closer to that determined by the pointer), to increase its precision (giving closer values when the procedure is repeated within a short period of time), to increase its stability (giving closer values when the procedure is repeated after a long period of time) or to increase its reliability (giving closer values when the procedure is repeated in different laboratories, by less skilled operators, in slightly different ways or using different equipment). The most plausible reason for discarding a pointer altogether, on this straightforward view, is that an alternative stipulation promises realizations of increased accuracy, precision, stability or reliability. Less drastically, a pointer itself may be honed in order to improve its realizations ("this bar being subject to standard atmospheric pressure and supported on two cylinders of at least one centimetre diameter, symmetrically placed in the same horizontal plane at a distance of 571 mm from each other" (CGPM, 1928, p. 49)). All of this is the work of the metrologist, who responds to demands from the rest of science (and beyond) for increased accuracy, precision, stability and reliability in scientific standards by improving existing pointers and their realizations, as well as developing new ones.

This straightforward view of scientific standards is one that makes sharp distinctions between what is being measured, the definition of a unit of that whatness and the method (or methods) by which it is realized. The distinctions are most easily played out by supposing that science divides neatly into three activities: theoreticians generate models of nature from the evidence before them; metrologists develop the measures required to test those models; experimentalists conduct the tests; theoreticians respond to the results with new models, etc. The metrological stage is, however, so often overlooked that of these three transitions, only that between the theoretician and experimentalist has been analysed in depth by philosophers. On the straightforward view of scientific standards, it is, in addition, taken for granted that metrology responds to (but does not directly inform) scientific theory: our understanding of our measures comes after and makes no contribution to our understanding of what we are measuring. A conceptual change driven by scientific theory can thus only be acknowledged in metrology by replacing a pointer ("Because we have a better understanding of what *length* is, we will now define a metre by the speed of light"). It is also the case that metrology provides for, but does not directly respond to, the work of experimentalists. On the straightforward view, an experiment can only begin once the metrological work required for that experiment has come to an end. The instruments used by the experimentalists are created and calibrated elsewhere. Although it has been appreciated generally that scientific activity, progress and knowledge cannot be neatly divided into the theoretical and experimental, the consequences for how we view the place of metrology in science have not perhaps been so thoroughly considered. The assumption of the isolation of metrology (that is does not contribute to scientific progress in profound ways) results from juxtaposing realism regarding the posits of science with a conventionalist about the standards of science; it is one that can occasionally be seen operating within the scientific community. For example, it is perhaps one of the reasons why it was difficult to recruit graduate scientists to metrological work in the 1960s (Quinn, 2011).

It is at least true that, outwardly, metrology does not appear to be a theoretical science in the same way as the physical sciences it supports; lacking its own models of nature, metrology stands apart. Even a cursory glance at the history of metrology furnishes us with evidence for the lag between theoretical world of science and its metrological afterthought: after the onslaught upon our concept of mass by twentieth-century physics, for example, the scientific community remained tied to realizing mass by the pointer that had been manufactured before Albert Einstein's birth and officially accepted in the early years of his childhood. On the straightforward view, how we define and measure a unit doesn't matter to our understanding of the quality that unit represents. The decisions regarding the scales against which we measure it – and thus the choice of pointers and methods of realization – are entirely matters of convention, for it is a choice guided by human values and not nature. The sharp distinction between the theoretical study of a quality and the conventional methods of measurement places metrology in its entirety in the "subjective", "arbitrary" or "conventional" bin. I call it *the conventionalist account of scientific standards.*

Despite the name, this viewpoint does not obviously sit astride a broader – a more generic – conventionalism about scientific concepts, of the kind associated with the likes of Ernst Mach, Henri Poincaré, Percy Bridgman, Rudolf Carnap or Hans Reichenbach. The conventionalist view of scientific standards does not capture the thinking of these more careful conventionalist thinkers and, as we shall see, is ironically in deep conflict with generic scientific conventionalism. It is, in fact, more neatly accommodated by a naïvely realist view of the physical sciences: a quantity sits in nature, regardless of whether and how we decide to measure it. (I appreciate that the label "conventionalist" is therefore confusing. However, I hope that my terminology will not be unfortunate in the long run, because it will encourage us to pick apart different threads of conventionalist thinking about science, not necessarily mutually supporting.) When lined up with naïve realism, as well as the assumption that the theoretical and the experimental can be cleanly distinguished in science, it no doubt appears that I have prepared the perfect straw-man for myself. However, I do not hold the view that the conventionalist account of scientific standards is always wildly inaccurate or always inapplicable. I do not deny its usefulness and – if pressed to talk about such things – I would admit that it contains an element of truth. I accept the account as one model, albeit an elementary and particularly unenlightening one, regarding the connection between scientific standards and physical theory. And I will be pointing out the advantages it has, when dealing with scientific standards, over more generic conventionalist accounts. The theses I present here may seem radical alternatives, but I believe that no account of scientific standards can ever be more than an iteration of the conventionalist's. My suggestion is not that these distinctions are to be done away with altogether, but that they should be seen for what they are, conventions themselves. My first intention in articulating the conventionalist account of scientific standards, then, is to highlight its prevalence in philosophical and more general thinking and our unwitting reliance upon it. It is easy – perhaps too easy – to interpret metrological history through its lens. As a result of its hold on our thinking, it leads both realists and anti-realists astray: having appreciated the theory-ladenness of experiment, there remains no interesting question to be asked regarding the role of scientific standards in physical theory. Despite, then, what I have already said regarding the resonance between the conventionalist account of scientific standards and naïve realism, I argue that it is also the entrenched view (albeit somewhat warped) within anti-realist philosophies. For example, it is because anti-realists did not more comprehensively reject the conventionalist account of scientific standards that the philosophical community was burdened with a doomed form of operationalism, limping before the race to explain the meaning of scientific terms began. Whether we have noticed it or not, the conventionalist account of scientific standards is with us, in all its damned simplicity. In Section 2, I present an overview of modern metrology and the current reform to the international system of scientific units (SI), demonstrating how natural it is, even for metrologists themselves, to assume the position of a conventionalist with regard to scientific standards.

My second intention is to pinpoint where the conventionalist account of scientific standards fails to account for some of the finer details of metrological practice and to therefore indicate what a *metrological account of scientific standards* might look like. I use the term generically to refer to any account that pays attention to such details. The BIPM defines a "measurement standard" as a, "realization of the definition of a given quantity, with stated quantity value and associated measurement uncertainty, used as a reference" (2012, p. 46). I use "scientific standard" because, as will become clear (Section 5), I view standards as the repeatable procedures, ratified by metrologists, that produce these realizations. If this thesis has a constructivist shimmer to it, it is complemented by the thesis that metrological ratification is tightly constrained by experimental realities (Section 3). I do not, however, defend a particular position here, but suggest three theses that a metrological account of scientific standards might incorporate. The case study at hand is the current redefinition of the kilogram. I describe here some of the details of this reform: the motivations that led to the change (Section 3); the development of the Kibble balance (Section 4); and the procedures for cleaning and washing kilogram prototypes (Section 5). The underlying theme to the analysis is to consider how the clean-line thinking of the scientific-standard conventionalist falls apart in the face of these details: between precision and accuracy (Section 3); between calibration and experiment (Section 4); and between meaning and method (Section 5). In each of these three cases, I explore an extreme response to the case study, before settling upon a weaker statement of it, in an attempt to develop a metrological account of scientific standards.

In Section 3, I argue that the desire for stability in scientific standards has a theoretical aspect that cannot merely be explained as the requirement to reduce the uncertainty of a standard in the long term. The motivation for the redefinition of the kilogram is both pragmatic and theoretical in nature. One conclusion available to the analyst is to reject the naïve realism associated with the conventionalist account of scientific standards, instead recognizing a milder and yet more specific realism. Roughly, *metrological realism* is the thesis that uncertainty of measurement does not just indicate the practical and technological limitations of empirical inquiry, but also measures the limits placed upon science by the way the world is. In attempting to reduce the uncertainty of measurement, then, metrologists are investigating and modelling nature.

In Section 4, I explain how the Kibble balance, once used to determine the Planck constant, will instead be used to realize the kilogram at the highest level. At a minimum, this indicates that the choice of metrology determines the lines between experiment and calibration. A stronger conclusion available to the analyst is to suppose that an experiment includes all the chains of calibration that are associated with its determinations. Giving calibration the same epistemic status as experimentation, and perhaps metrology the same as other physical sciences, I propose the *metrological unification of the physical sciences*. Roughly, this is the view of the physical sciences as the collection of all physical determinations, whether of experiment or of calibration, together with the theoretical associations of these determinations.

In Section 5, I describe the current official procedure for cleaning and washing kilogram prototypes. The study indicates that there is no distinction between a pointer (the definition of a standard) and its *mise en pratique* (its official method of realization) that is of epistemic importance. This paves the way for considering that a *mise en pratique* contributes to the meaning of a scientific standard. To put the point a little too provocatively: the way we measure the world contributes to the meaning of science. The point leads us to consider operationalism and again raises the problem of accommodating metrological progress from a conventionalist point of view. In response to this, I develop a more thoroughgoing, but yet much weaker version, of Bridgman's operational attitude. Roughly, *metrological operationalism* is the thesis that measurement procedures contribute, at least partially, significance to scientific standards and the quantities associated with these standards.

Altogether, this points towards an (as yet hazy) metrological account of scientific standards. Metrology is more theoretical, empirical and meaningful than the conventionalist account of scientific standards allows. I speculate about the connections between these theses in Section 6. Regardless of the details of my suggestions, however, I hope that the question regarding the role of scientific standards in physical theory now presents itself as an interesting one. To begin the analysis, I must describe the deep hold the conventionalist account of scientific standards has on the story of metrology.

2   A conventionalist perspective upon the history of modern metrology

There are good reasons to take the modern age of metrology to begin in 1875, when representatives from 17 nations met in Paris to sign a treaty. The Metre Convention created the political structure through which international agreement on matters of scientific standards has since been made: it brought about the International Bureau of Weights and Measures (BIPM), an international research organization working to improve scientific standards, and its governing body, the General Conference on Weights and Measures (CGPM), a quadrennial meeting of delegates from member governments that directs the research at the BIPM, as well as the International Committee for Weights and Measures (CIPM), 18 individuals from member states, who make recommendations to the CGPM. The metric system itself dates back to revolutionary France of the 1790s, when the metre was defined by a portion of the Earth's circumference and the kilogram was defined by a volume of water at its densest. With these definitions, however, the original creators of the metric system had intended to establish natural scientific standards, beyond the reach of human error and without the need of human maintenance.[1] Although the original creators expected that it might be one day necessary to replace the platinum-iridium artefacts that exemplified the new standards, they believed that the real work of length and mass metrology had been completed. The nineteenth century witnessed the growing recognition that the metric project had failed in this intention; the understanding that scientific standards develop alongside the rest of science brought about the founding of the BIPM and metrology as we know it today.

As the name suggests, the original purpose of the Metre Convention was to advance length metrology. The International Metre Commission, which led directly to the Metre Convention, was assembled in 1869 in response to the difficulties faced by the international geodesic community in establishing agreement and uniformity in length measurement.[2] In addition, it was appreciated that a more precise calibration of length could be obtained against a standard defined by two fine lines marked upon a metal bar, instead of one defined by the entire length of a bar. The French section of the commission initially understood their remit to be limited to the creation and distribution of replicas of the original metre prototype, both end- and line-standards. The project was soon shaped, however, by the international view that the original artefact was to be replaced. And because there was a metric connection between mass and length – the kilogram had originally been defined by a *decimetre* cube of water at 4°C and the original kilogram prototype had been made at the same time as that of the metre – the commission nervously took responsibility for creating a new kilogram artefact at the same time. Doubts were expressed that they had the power to do so, yet members of the commission voted, by the narrow margin of ten votes to eight, to take on this additional project. The practical demands of international geodesy brought about the creation of new prototypes of both the metre and the kilogram.

Modern metrology was thus born in a whirlwind of conventionalism, acknowledging that scientific standards were necessarily designed, made and looked after by metrologists according to the demands of the scientific community. And the conventionalist viewpoint was not restricted to the metrological community, but was reflected and a reflection of the conventionalist views of physicists and philosophers at this time. In the period 1875–1930, metrology held a high status within both physics and philosophy of physics. Metrological work attracted the attention of successful physicists in the UK (including James Clark Maxwell, Lord Kelvin, and J. J. Thomson) and was rewarded with Nobel Prizes (Alfred Michelson in 1907 and Charles Guillaume, director of the BIPM, in 1920).[3] The philosophy of the period was informed by metrological practice (before his rise as a philosopher, for example, C. S. Pierce had contributed to measurements of the intensity of the Earth's gravitational field). The rising philosophies of science, whether labelled pragmatic, logical, empirical or conventional, paid attention to the conventional nature of metrology (most notably, I will be turning to the work of Reichenbach in Section 4).

As the twentieth century progressed, one of the more salient features of metrological progress was an increasing rigour, arising not least as a result of the work of the BIPM and reflected in increasing standardization, efficiency and formality. Developments of this kind are readily interpreted from the conventionalist's point of view. In my first example of this interpretation, consider that, during the twentieth century, it became standard metrological practice to associate each measurement in the physical sciences with a parameter that "characterizes the dispersion of the values that could reasonably be attributed to the measurand" (BIPM, 2008, p. 2). The estimation of uncertainty is not entirely algorithmic. It does include a statistical analysis of the experimental data and therefore takes into account the variation of data points. (Thus uncertainty is, strictly speaking, neither attached to a measurement procedure in general, nor to a single enactment of it; uncertainty is, in the first place, associated with a limited number of repetitions of a procedure, which together constitute one measurement.) Uncertainties calculated statistically are called "Type A" uncertainties; in addition, estimates are made of non-statistical, "Type B" uncertainties, which result from other sources, including calibration certificates, manufacturer's specifications and from common sense (Bell, 2001, p. 11). A *combined standard uncertainty* is calculated from the (potentially many) Type A and Type B uncertainties (BIPM, 2008, p. 7). The conventionalist account of scientific standards encourages us to view metrological progress, including the current reform to the SI, as an attempt to decrease the uncertainty with which we can make measurements and thus to view the combined standard uncertainty as a measure of the imprecision, instability, unreliability and inaccuracy of measurement. Most obviously, it is a measure that, in its inclusion of Type A uncertainties, relates directly to the imprecision with which a standard has been realized. In theory, statistical analysis of data taken from different laboratories and across time periods can also reflect the unreliability and instability of a standard's realization. In the case of the IPK, metrologists have attempted to quantify the uncertainty associated with its instability; the conventionalist account of scientific standards appears to be supported by the metrologists who reason that the last artefact from the SI must be given up because of its "inherent instability" (Davies, 2005, p. 2263). Further indicators of unreliability are represented by Type B uncertainties, including, for example, the calibration of instruments used in the measurement and dependencies upon external experimental values. The demand for an accurate standard, on the other hand, is dealt with in a less direct way by combined standard uncertainty. It is necessary to infer the accuracy of a measurement from the coherence of results obtained using methods that are theoretically different; the combined standard uncertainty merely provides a range of values from which it can be determined whether results are compliant.

A second example of interpreting the twentieth century metrology through the conventionalist lens comes from considering the increasing formality of metrological practice during this time. The BIPM came to recognize its responsibilities for determining and communicating each *mise en pratique*, a procedure by which an SI standard is realized "at the highest level" (CIPM, 2008, p. 62). The term applies to the procedures for realizing a standard using an artefact (regarding the handling of the IPK, for example) and to those for realizing a standard from a fundamental constant (regarding the operation of the Kibble balance, for example). The BIPM now takes great care when choosing the wording of a *mise en pratique*, as well as that of the definition of a standard, and distinguishes between the two. It recognizes, in fact, that the distinction has been made especially sharp by the current reform of the SI, in which the kilogram, the ampere, the kelvin and the mole are being redefined by fixing the values of four physical constants (the Planck constant, the elementary charge, the Boltzmann constant and the Avogadro constant, respectively), in addition to the three physical constants whose numerical values have been fixed to date (the caesium hyperfine frequency, the speed of light in vacuum and the luminous efficacy of a defined radiation). Thus, at the core of the reformed SI, are fixed values for seven physical constants of nature, which correspond to seven standards of measurement (given in Table 7.1). In marked contrast to these seven numbers, are the methods being developed that will stand as *mise en pratiques* for the seven associated units. Furthermore, the BIPM (2013, pp. 9–10) makes the distinction precisely for the reason of reducing uncertainty in the long term:

> The use of seven defining constants is the simplest and most fundamental way to define the SI. . . . In this way no distinction is made between base units and derived units; all units are simply described as SI units. This also effectively decouples the definition and practical realization of the units. While the definitions may remain unchanged over a long period of time, the practical realizations can be established by many different experiments, including totally new experiments not yet devised. This allows for more rigorous intercomparisons of the practical realizations and a lower uncertainty, as the technologies evolve.

Beneath some statements of the BIPM, regarding its intentions and its purpose, then, it is possible to interpret a conventionalist view of scientific statements. Metrologists redefine pointers, as well as the realizations for those pointers, in order to increase the precision, reliability, stability (and perhaps accuracy) of scientific standards. The single aim of metrology is to reduce the uncertainty with which measurements can be made. On a first pass, modern metrology – including the current reform of the SI – conforms to the conventionalist account of scientific standards.

Table 7.1 Currently accepted values of the physical constants that will be fixed by the latest reform of the SI

Physical constant	2010 CODATA value
the unperturbed ground state hyperfine splitting frequency of the caesium 133 atom $(_{133}Cs)_{hfs}$	9,192,631,770 hertz
the speed of light in vacuum c	299,792,458 metres per second
the Planck constant h	$6.626,069,57 \times 10^{34}$ joule second
the elementary charge e	$1.602,176,565 \times 10^{19}$ coulomb
the Boltzmann constant k	$1.380,648,8 \times 10^{23}$ joule per kelvin
the Avogadro constant N_A	$6.022,141,29 \times 10^{23}$ reciprocal mole
the luminous efficacy K_{cd} of monochromatic radiation of frequency 540×10^{12} hertz	683 lumen per watt

Source: BIPM, 2013

3   Motivations for a new definition of the kilogram and the case for metrological realism

In 1889, the original exemplar of the kilogram was replaced by a platinum-iridium artefact manufactured in London by George Matthey. The International Prototype Kilogram (IPK) was accepted as the kilogram standard at the first meeting of the CGPM (1890, p. 34): "This prototype shall henceforth be considered as the unit of mass". More than anything else, what marked the new kilogram standard from the Kilogram of the Archives was that it was one of 42 prototypes, all made to the same specifications. In fact, the creation of the new metric standards under the direction of the newly-established BIPM had been delayed during the period 1774–1882 because of the desire to make all the prototypes from a single casting of platinum-iridium, a requirement that was eventually dropped (Quinn, 2011, Ch. 4). The commissioners believed it of importance that the prototypes be as similar as possible because they were aware that, in creating many prototypes of the new standards, they were creating a check upon the stability of the IPK. It was envisaged that the 42 prototypes would be brought together at intervals to recalibrate them against the IPK. The first comparisons were performed before the prototypes were distributed to governments worldwide, in the period 1886–1889, by Max Thiesen and Damian Kreichgauer. Each prototype was compared directly with the IPK, as well as with 12 other prototypes. From a statistical analysis of the variations in the measurements taken, it was concluded that the mass of each prototype with respect to the IPK was accurate to within 0.002 mg (Thiesen, 1898, Ch. 17). As we have seen, the concept of uncertainty has since developed. This analysis would thus not be accepted today: consideration of additional causes of uncertainty would result in a higher measure (Quinn, 2011, p. 123).

The second periodic verification of the national prototypes was conducted by Albert Bonhoure, in the period 1946–1953 (CIPM, 1946, pp. 171–178). This included additional prototypes, numbered between 44 and 55 (non-inclusively), made by Matthey's firm as the membership of the Metre Convention increased in the first half of the nineteenth century. The third periodic verification of the national prototypes was conducted by Georges Girard, in the period 1988–1992 (CIPM, 1993, pp. G35–G50). Again, the collection of prototypes had expanded; the new additions were numbered between 56 and 65 (non-inclusively). Altogether, the third verification measured the mass of 51 kilogram prototypes with respect to the IPK (listed in Table 7.2): the six official copies of the IPK (K1, No. 1, No. 7, No. 8(41), No. 32, No. 37 and No. 38), the three working copies of the IPK belonging to the BIPM (No. 9, No. 25 and No. 31), and 42 national prototypes (most of which were held by governmental metrological laboratories).[4]

In all, the third verification confirmed what had already been showing in the results of the second: in 1992, the mass of an original copy of the IPK was, on average, 25 μg more than it was in 1889 (shown in Figure 7.1); the mass of a secondary copy of the IPK was, on average, 40 μg more than it was in 1946 (shown in Figure 7.2). One way of interpreting this data is to suppose that the mass of the IPK is itself drifting, losing mass at a rate of the order of 0.5 μg per annum.

Table 7.2 Results of the third periodic verification of the national prototypes.

International Prototypes			茶 1 kg		
Official Copies	Kl	1 kg + 0.135 mg		No. 32	1 kg + 0.139 mg
	No. 7	1 kg − 0.481 mg		No. 43	1 kg + 0.330 mg
	No. 8(41)	1 kg + 0.321 mg		No. 47	1 kg + 0.403 mg
BIPM prototypes			No. 25	1 kg + 0.158 mg	
			No. 9	1 kg + 0.312 mg	
			No. 31	1 kg + 0.131 mg	
National and other prototypes	No. 2	Romania			1 kg − 1.127 mg
	No. 5	Italy			1 kg + 0.064 mg
	No. 6	Japan			1 kg + 0.176 mg
	No. 12	Russian Federation			1 kg + 0.100 mg
	No. 16	Hungary			1 kg + 0.012 mg
	No. 18	United Kingdom			1 kg + 0.053 mg
	No. 20	United States of America			1 kg − 0.021 mg
	No. 21	Mexico			1 kg + 0.068 mg
	No. 23	Finland			1 kg + 0.193 mg
	No. 24	Spain			1 kg − 0.146 mg
	No. 34	Académie des Sciences de Paris			1 kg − 0.051 mg
	No. 35	France			1 kg + 0.189 mg
	No. 36	Norway			1 kg + 0.206 mg
	No. 37	Belgium			1 kg + 0.258 mg
	No. 38	Switzerland			1 kg + 0.242 mg
	No. 39	Rep. of Korea			1 kg − 0.783 mg
	No. 40	Sweden			1 kg − 0.035 mg
	No. 44	Australia			1 kg + 0.287 mg
	No. 46	Indonesia			1 kg + 0.321 mg
	No. 48	Denmark			1 kg + 0.l 12 mg
	No. 49	Austria			1 kg − 0.271 mg
	No. 50	Canada			1 kg − 0.111 mg
	No. 51	Poland			1 kg + 0.227 mg
	No. 53	Netherlands			1 kg + 0.121 mg
	No. 54	Turkey			1 kg + 0.203 mg
	No. 55	Fed. Rep. of Germany			1 kg + 0.252 mg
	No. 56	South Africa			1 kg + 0.240 mg
	No. 57	India			1 kg − 0.036 mg
	No. 58	Egypt			1 kg − 0.120 mg
	No. 60	People's Rep. of China			1 kg + 0.295 mg
	No. 65	Slovak Rep.			1 kg + 0.208 mg
	No. 66	Brazil			1 kg + 0.135 mg
	No. 68	Dem. People's Rep. of Korea			1 kg + 0.365 mg
	No. 69	Portugal			1 kg + 0.207 mg
	No. 70	Fed. Rep. of Germany			1 kg − 0.236 mg
	No. 71	Israel			1 kg + 0.372 mg
	No. 72	Rep. of Korea			1 kg + 0.446 mg
	No. 74	Canada			1 kg + 0.446 mg
	No. 75	Hong Kong			1 kg + 0.132 mg
	No. 3	Spain			1 kg + 0.077 mg
	No. 62	Italy (IMGC)			1 kg − 0.907 mg
	No. 64	People's Rep. of China			1 kg + 0.251 mg

Source: CIPM, 1993, G43

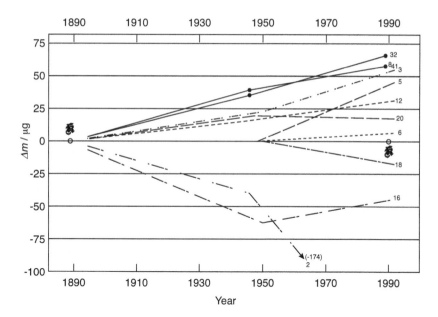

Figure 7.1 Change in mass of the national prototypes No. 2 to 20 (those made prior to the first verification of 1889), as well as official copies No. 8(41) and No. 32, with respect to the IPK (CIPM, 1993, G45).

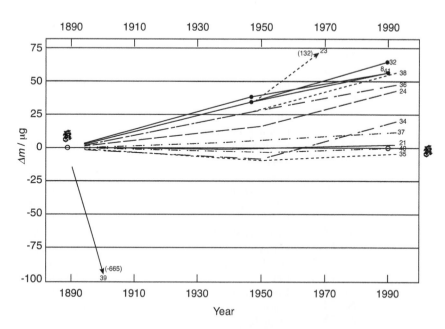

Figure 7.2 Change in mass of the national prototypes No. 44 to 55 (those made after the first verification of 1889 but before the second of 1946), as well as official copies No. 8(41) and No. 32, with respect to the IPK (CIPM, 1993, G45).

The metrological community itself, however, has shown some caution in reaching this conclusion. Metrologists appreciate that their current knowledge of the behaviour of platinum-iridium prototypes is too limited to determine the causes of the apparent drift (Quinn, 2011). It is plausible, for example, that the discrepancies between the IPK and the national prototype are due to the response of the IPK to washing and cleaning: Girard's tests demonstrated that the IPK changed in mass more markedly after washing and cleaning than other prototypes (shown in Figure 7.4).[5] What is clear is that, over long periods of time, the prototypes are unstable to an extent that can be detected.

It was not, however, this result that prompted the current reform of the mass standard. The apparent drift of 0.5 μg per annum in the kilogram standard has not yet created any practical problems for commercial, industrial or scientific users of the kilogram. (To put the drift into perspective, consider that Girard (1990) judged that the mass of each national prototype had been measured against the IPK with an uncertainty of 2.3 μg; such mass calibrations are the most certain that can be done in SI units because all other mass measurements must take into account this uncertainty.) Furthermore, it has long been presumed by metrologists that the mass of a metal piece changes over time; the result of the third verification of the IPK held little surprise. If the drift had not been apparent, the desire to replace the IPK would remain. Indeed, the desire to replace an artefact mass standard with something "more fundamental" existed long before the third – and even the second – verification took place. The third periodic verification of the national prototypes was conducted, in fact, because the metrological community could not achieve what it really wished to. The 15th CGPM of 1975 requested that the BIPM conduct the verification, as well as continue research into the improvement of mass standards comparisons, for the reason, "that there is no immediate prospect of defining the mass unit in terms of atomic constants with a comparable precision" (CGPM, 1975, pp. 103–104).

The desire to define a mass unit in more fundamental terms than by a metal artefact is older than the technological ability and the practical requirements to do so. It is a sentiment that echoes throughout the metrological literature: in the papers and review articles of *Metrologia* since the journal's inception in 1965; in the annals of the CGPM, produced every four years since 1889 and the associated minutes of the CIPM's yearly meetings since 1876; in the parliamentary archives of the French Revolution concerning the creation of the original metric system; and in the works of natural philosophers of the seventeenth and eighteenth centuries.[6] The Metric Convention turned out to be only a temporary postponement of this goal. The concept of a fundamental scientific standard is one that has never gone away entirely, although it has changed in this time, most obviously in response to a changing body of scientific knowledge. A fundamental standard has usually been understood to be in some way given or representative of nature (an idea that has been associated with immutability, uniformity, self-maintenance, reproducibility, deliverance by experiment and explainability to aliens).[7] Since the mid-nineteenth century, however, it has increasingly been understood to be a molecular or atomic standard, one that relies upon counting phenomena in the microscopic realm. One of the earliest and clearest articulations of this point was made by James Clerk Maxwell (1870, p. 7):

> If, then, we wish to obtain standards of length, time, and mass which shall be absolutely permanent, we must seek them not in the dimensions, or the motion, or the mass of our planet, but in the wave-length, the period of vibration, and the absolute mass of these imperishable and unalterable and perfectly similar molecules.

By the turn of the twentieth century, delivering fundamentality via atomic standards had become a possibility. This was recognized by the astronomer David Gill in his presidential address to the British Association for the Advancement of Science in 1907. Gill acknowledged that he had been influenced by Maxwell's dream of a standard communicable to aliens during an 1859 lecture Maxwell gave in Aberdeen. He also recognized the very practical issue of breaking or losing an artefact standard. It was scientific discovery that had seemingly revealed the answer to these issues, nature herself directing metrologists towards an atomic metrology. Thus, Gill (1907, p. 195) explained that the International Prototype Metre was not scientifically described by the label "one metre" but only

> as a piece of metal whose length at 0° C. at the epoch A.D. 1906 is 1,553,164 times the wavelength of the red line of the spectrum of cadmium when the latter is observed in dry air at the temperature of 150 °C. of the normal hydrogen-scale at a pressure of 760 mm. of mercury at 0° C.

The description made use of the most recent determinations of the red-light wavelength emitted from cadmium, by Alfred Pérot and Charles Fabry in 1906.[8] From Gill's point of view, there was a better (a more scientific) description of "one metre" than that given by the International Prototype Metre. It has not solely been a desire to decrease the uncertainty with which artefacts can be compared that has driven SI reform, for the challenge comes from the opposite direction. Since the creation of the IPK, metrologists have wanted instead to decrease the uncertainties with which a mass standard could be realized in alternative, more fundamental ways. The prospect of doing so eventually came from two independent lines of development. On the one hand, Bryan Kibble developed a balance that compared gravitational with electrical power; developments in quantum electrodynamics then enabled a theoretical link to be made between a gravitational mass and quantum constants. On the other hand, computing technologies brought the possibility of measuring mass by counting silicon atoms. In both cases, metrologists at the turn of the twenty-first century were working to improve the uncertainties of these techniques, aiming only to match the uncertainty associated with calibrating against the IPK, and not to improve upon it. The target uncertainty was accepted to be 2 in 10^8 parts, a measurement of a kilogram to within 20 μg (Kelley, 2001, p. 860).

The reform to the kilogram was not, then, immediately driven by a desire to decrease the uncertainty with which platinum-iridium prototypes can be compared. The most obvious first responses to the apparent drift of the IPK with respect to its copies, after all, is to investigate the causes of the drift (a line of research that has been left relatively unexplored by the metrological community) or to define the kilogram by the average mass of a collection of pieces (a possibility that has not seriously been considered since the installation of the IPK).[9] The progress of metrology is occasionally driven by more theoretical concerns, which can be interpreted as an expression of a scientific realism of sorts. The apparent drift between the IPK and its prototypes is, after all, only a relatively minor part of a larger problem regarding the long-term stability of platinum-iridium artefacts. There remains the further possibility of a more unsettling drift, in which all the platinum-iridium pieces are drifting from their original masses. The metrological community accepts that the masses of the prototypes changes in the long term and estimates this change (without experimental evidence) to be within ten times that of the drift between the pieces themselves, resulting in an upper limit of 5 μg per year (Quinn, 2011). Such estimates assume that the mass of the iridio-platinum prototypes is to be compared to a more accurate indicator of mass. There is, however, no experimental warrant for believing that the Planck constant will provide more stability to mass measurement in the longer term. The turn towards the fundamental constants is founded on Gill's assumption, seemingly both innocuous and true: the natural world (and not just the practical requirements and realities of experimental science) determines better and worse ways of defining scientific standards.

Conventionalism regarding scientific standards is naturally coupled with a naïve realism regarding the posits of scientific theory and can, in this way, account for the theoretical motivations of metrological progress. It does this, however, with a heavy hand. There is the temptation to go further than is strictly warranted by the details of metrological practice and assume the existence of fundamental standards in nature. In comparison, the realism called for by Gill's assumption makes only mild ontological commitments. Furthermore, because the conventional work of the metrologist is crudely aligned with the investigative work of the theoretician, it is assumed that the kilogram is only reformed once a new understanding of mass is delivered as a result of progress in the physical sciences. The suggestion to fix the value of the Planck constant was, however, one that emerged within metrology. The thesis that the Planck constant is invariable across space and time is not one that is thoroughly integrated into theoretical physics: the Copernican principle that the laws of physics are the same for all observers is compatible with smooth changes in the fundamental constants and physicists continue to speculate whether the fundamental constants change over time. The point is most famously expressed by Dirac (1937), but remains a part of contemporary physics (Avetissian, 2009). It remained for metrologists to test the stability of the Planck constant and perform the precision measurements required to demonstrate it provided a suitable grounding for a scientific standard (Steiner et al., 2005; Eichenberger et al., 2009). It was not because science had progressed to the stage that the uncertainty of the Planck constant was on a par with mass measurements that the SI reform was initiated, but the reverse: it was metrologists who asked how well the constant is known and determined to make further precision measurements of it. My claim here is not just that the line between calibration and experiment blurs, but so too between metrology and precision measurement, often considered a part of the physical sciences proper. The experimental and theoretical work of determining the constancy of the Planck constant is not done separately from metrology.

Furthermore, an alternative proposal for defining the kilogram, involving the counting of silicon atoms, was rejected, despite it being recognized as closer to our intuitive understanding of what mass is and thus "more readily comprehensible" (Hill, Miller, & Censullo, 2011). We shall see that the Kibble balance weighs masses using electromagnetic units, but there is no drive from theoretical physics that encourages metrologists to suppose that the true essence of mass is explained by electromagnetic theory. It was for the metrological community to determine how to interpret mass for the purposes of measurement. Ironically, it relies upon the quantification of precision and stability in order to do this: these measures are not merely indicators of the practical difficulties faced by the users of the kilogram, but indicators of what nature presents as a better measure of reality.

The metrological investigation of the invariance of the Planck constant and of material artefacts is only just beginning. As part of the current reform, metrologists are currently creating a collection of kilogram artefacts to replace the national prototypes. It was important, when making the last collection, that they were as similar as possible. It is important, in the current work, to include kilograms of many different materials. The purpose will be to monitor the stability of artefacts against the Planck constant. It is plausible that there are surprises are in store for us regarding the stability of mass measurements and, as a result, our understanding of how best to represent mass. But in any case, the responsibility for this lies firmly in the hands of the metrological community. That responsibility includes determining the status of the statement, "The Planck constant does not change in space or time", and its role in physical theory. Metrologists have the power to underline it as an important theoretical principle, or leave it as an empirical possibility, but it remains open to the results of precision testing. It won't do, then, to consider the metrological choice as entirely or merely conventional. In comparison to the realism that is usually coupled with scientific-standard conventionalism, the realism called for by Gill's assumption makes strong epistemological demands.

The motivation for the stability of a standard is thus a "thick" one: as well as being a practical desire that our standards do not change over time, it is an indication that we are measuring the right thing.[10] From the latter viewpoint, precision is not just a useful thing to have, but it is a mark that we are closing in on the true regularities of nature. Metrologists are not merely responding to physical theory in deciding to redefine the kilogram by fixing the value of the Planck constant, for metrology is concerned with revealing nature as she is, determining by scientific experiment what is the most uniform and immutable in nature. Although the most obvious sign of that a physical determination is accurate comes from coherence between two different methods of measuring it, it is also the case that precision indicates that a measure is – as metrologists tend to say – "more fundamental" than another (Mills, Mohr, Quinn, Taylor, & Williams, 2011, p. 3). One way of acknowledging the dual aspect of the motivation behind the current SI reform is to accept that metrological progress is best captured by what is both a mild and a strong realism. Towards this end I propose the thesis of *metrological realism*: nature does not endorse all scientific standards equally; some scientific standards have the potential to be realized with lower uncertainties because they more closely reflect the regularities of nature.

Analysis of the theoretical desires behind the reform of the SI thus reveals, I believe, that the development of scientific standards is more closely associated with physical theory than the conventionalist account allows. There is no doubt that, in choosing a certain number, there is a conventional element to fixing the value of the Planck constant to that number. But the conventional aspect of this unit-setting is so remarkably tame, no more than is to be found elsewhere in the physical sciences, that it won't do to write of the kilogram itself as a convention. There is more to be said about the nature of this theoretical aspect of metrology, however. In the next section, I go on to consider two metrological procedures associated with the IPK: the current BIPM procedure for cleaning and washing of kilogram prototypes and the proposed use of the Kibble balance to calibrate a mass standard. So far, I have been challenging the conventionalist distinction between accuracy and precision; I now look to take the challenge to the distinction between experiment and calibration and between a pointer and its *mise en pratique*.

4   The Kibble balance and the metrological unification of the physical sciences

In 1976, Kibble (1976) proposed that a beam balance could be used to compare a gravitational power with an electrical one; the principle of the resulting Kibble balance is illustrated in Figure 7.3. A mass m is suspended from one side of the balance; an electric coil of length L is suspended from the other in radial magnetic field of flux density B. To begin with, the balance is brought to equilibrium by passing a current I through the coil. The beam balances when the electromagnetic force BIL exerted upon one side of the beam matches the gravitational force mg upon the other:

$$mg = BIL$$

In the second phase of the procedure, the coil is moved downwards with velocity v, which induces an electric potential U, where $U = BL\,v$. Thus, the mass suspended from the balance can be expressed as a voltage and a current:

$$mg\,v = IU$$

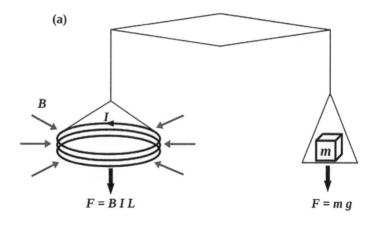

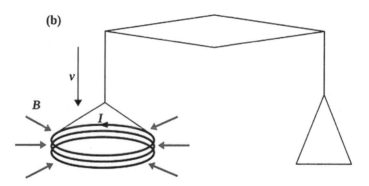

Figure 7.3 The principle of the Kibble balance: (a) when in equilibrium; (b) when in motion.

Or, replacing the current I with U/R (by Ohm's law), where R is the resistance of the coil:

$$mg\,v = U^2/R \qquad (1)$$

This enables a mass measurement to be taken by reading an electromagnetic force. If U, R, g and v are measured in SI units, the resulting mass will be given in kilograms. A thorough description of the procedure is given by Kibble and Robinson (2003).

At first, the Kibble balance was proposed as an improvement to the ampere balance, and it did not have the promise of an atomic standard. It was the confirmation of the quantum Hall effect in the 1980s that brought this about. The Hall effect is the presence of a transverse voltage in an electrical conductor. Quantized, a resistance (the transverse resistance of the Hall effect) is expressed in terms of the Planck constant h, the elementary charge e and an integer p:

$$R = h/pe^2 \qquad (2)$$

This allows for the resistance R of (1) to be expressed using microscopic quantities. The same can be done for the electric potential U of (1) by applying the theory of the Josephson effect, known since the 1960s. An electric potential exists at the junction of two superconductors separated by a thin layer of non-superconducting material and can be expressed in terms of the Planck constant, a frequency f (the phase difference between the metals), an integer q and the elementary charge e:

$$U = qfh/2e \qquad (3)$$

Thus a link can be made between a macroscopic mass and the Planck constant (inserting (2) and (3) into (1)):

$$m = q^2 f^2 he/4gv$$

As a result of advances in quantum electromagnetism, the Kibble balance now provided a new experimental way to determine the Planck constant, by suspending a known mass upon it. But it also offered the possibility that, if the numerical value of h were fixed, it could be used to assign mass measurements.

Curiously, then, an experiment that was once used to determine the Planck constant will, after the current reform of the SI, act instead as a procedure to calibrate masses against the new definition of the kilogram. Thus, the choice of a metrology determines whether a laboratory procedure (the operation of the Kibble balance) is to be considered an experiment (to determine a physical constant) or a calibration (to determine the mass of an intermediate standard). This somewhat challenges the assumption that it is scientific experiments that provide science with its empirical content, in contrast to calibrations, which are mere translations of that content into a more convenient, universal language. A calibration is, after all, usually understood as a way of translating the indications given from laboratory equipment into a language understandable to those outside of that particular laboratory. As the BIPM defines it (2012, p. 28):

> [A calibration is] an operation that, under specified conditions, in a first step, establishes a relation between the quantity values with measurement uncertainties provided by measurement standards and corresponding indications with associated measurement uncertainties and, in a second step, uses this information to establish a relation for obtaining a measurement result from an indication.

Without denying the physical necessity of talking in kilograms instead of the degrees of the pointer of the red balance in the Mott Building, the conventionalist account of scientific standards marks the moment when the position of the red-balance pointer was recorded as bringing the empirical part of the experiment to an end with respect to the mass measurement. Each experiment in the physical sciences is supported by a collection of calibrations, which are each supported by further calibrations, ultimately leading to one or more of the *mise en pratiques* of the SI standards, in a way that can be depicted by a spider-web diagram. From this viewpoint, it comes as a surprise that a reform of the SI units does not merely consist of supporting an experiment with alternative calibration chains, but redefines what counts as a calibration in the first place. It remains open, however, how deeply to interpret this observation and how far the conventionalist account of scientific standards must be altered in order to accommodate it, as is attested by the history of twentieth-century analytic philosophy, for this is, of course, a striking example of an old issue.

Logical empiricists saw the crash between experiment and calibration most clearly from the opposite side of the road. They appreciated that scientific knowledge previously understood to be purely empirical included a certain element of convention. Most famously, they acknowledged that, in order to formulate special relativity, Einstein had assumed that when light makes a return trip, it travels at the same speed on both legs of the journey. Reichenbach, amongst others, understood this assumption to be a convention, what he called a "coordinative definition" (Reichenbach, 1928). He considered such definitions to be logically necessary in order to allow the empirical work of science to get underway. Taking this route, it is possible to interpret the new use of the Kibble balance as altering only the linguistic framework in which science is conducted (Carnap, 1950). The statement "The Planck constant is $6.626,069,57 \times 10^{34}$ joule second" is empirical in one metrology and definitional in another. In accepting this, the logical empiricists recognized a single, one-way reliance between the conventional and the empirical.

It would, of course, be surprising, having rejected logical empiricism more generally in philosophy of science, to accept it as a success for metrology. And with hindsight, the first acknowledgement regarding the interplay between the conventional and the empirical is only the beginning of a slippery slope. I leave it open, however, how far a Quinean turn needs to be taken at this point. My sense is only that, after the toppling of logical empiricism, later philosophy of science has not been articulated with metrology in mind. It remains to test holistic theses in a metrological setting. Curiously, in this specific context, the associations between experiments marked by calibrations can be distinguished and separated from the larger web of knowledge, and, furthermore, translation occurs here *within* this web. It raises the question, then: what exactly is a calibration? The case for a Quinean metrology is yet to be made, for it is apparent that the analytic-synthetic distinction still remains deeply imbedded in thinking about measurement. Quine's philosophy is yet to be applied to the science of measurement and brought to bear upon the specifics of metrological practice. One suggestion in that regard is to indeed consider a more contained holism than Quine's, which can therefore recognize the diversity of the sciences and the compartmentalization of knowledge more generally. In this regard, I propose that it may sometimes be useful to consider the physical sciences as the collection of all determinations of physical quantities, regardless of whether they are, at any particular time, labelled as calibrations or experiments, and which are associated (in a theory-laden way), because some determinations make use of others. The metrological unification of the physical sciences gives metrology a similar epistemic status to other sciences, viewing the totality of determinations as providing the physical sciences with its empirical content. My sense is that the unification of the physical sciences is worth emphasizing today because of the philosophical community's growing understanding of the diversity of the sciences in general. I am not arguing against the seemingly obvious truth that the physical sciences are united by a number of natural laws that connect variables representing different kinds of quantity, but merely emphasizing that metrology is not merely an effort to express those equations with units, but it takes its part in determining which connections are to be regarded as definitional, and which hold empirical content. My thesis is not that metrology is everything in the physical sciences, but that it is more than current philosophical thinking allows for.

Although I do not suspect that a holism about language in general will prove very useful to the philosophy of measurement, it remains the case that if we wish to interpret the status of the Kibble balance experiment from a conventionalist viewpoint, we find ourselves between a rock and a hard place. The conventionalism of the logical empiricists diverges from the conventionalist account of scientific standards in two important ways. Although Reichenbach would agree that the choice of a scientific standard is a convention, he has reversed the order of communications between the metrologist and the physical scientist. On the logical empiricist account, coordinative definitions are necessarily prior to physical research; on the conventionalist account of scientific standards, as I have presented it, metrological pointers are handy aids for experimenters but are necessarily informed by physical theory. The scientific-standard conventionalist more closely describes the case of the kilogram reform here, given its dependence upon quantum electrodynamics. It is widely recognized, of course, that the logical empiricists failed to account for the extent with which observation is laden with theory. The trouble with Reichenbach's coordinative definitions is that they are improved by empirical research. Second, although it is not explicit in Reichenbach's discussion of coordinative definitions, it is reasonable to interpret his view as holding little distinction between a scientific standard's pointer and its method of measurement. In the next section (Section 5), we shall see that it is the scientific-standard conventionalist who is at a disadvantage here.

Both theories do best where they make small steps to allow for the interplay between the empirical and the conventional, the theoretical and the empirical, the work of the metrologist and that of the physical scientist. There is room on the philosophical landscape to draw up a metrological account of scientific standards that more keenly appreciates the depth of this interplay, and yet which perhaps does not discard these distinctions altogether. I take it that the most important steps in this direction have been made by Hasok Chang in his detailed historical accounts of progress in thermometry (2004). Each of Chang's case studies starkly demonstrates the empiricist's difficulty of making metrological progress because, on the face of it, there is nothing better to test our best scientific standards against but those standards themselves. Steering between conventionalism regarding scientific standards and operationalism, Chang outlines his theory of epistemic iteration, a version of coherentism, to account for how metrologists do, in practice, overcome this apparent circularity. Unlike the interval scales of early thermometry, the mass scale is a ratio scale, requiring (at least theoretically) just one fixed point. The underlying difficulty remains, however, that we wish to experimentally determine whether our chosen point is truly stable. Thus, the case study presented here is of the same ilk: how can we experimentally determine the most reliable measure of mass, without already having the most reliable measure of mass at hand? In today's reform of the kilogram, the solution has been partly provided by theoretical physics. It is partly on account of Chang's work that I reject the naïvely realist viewpoint that supposes metrology can rest on a foundation given by the constants of nature. It remains, after all, for metrologists to perform further precise measurements, testing and redefining scientific standards in the future. I have argued, however, that the coherentist has to accept a certain amount of realism in order to account for the intricacies of metrological history. I go on now to argue that, similarly, there is no need to reject all the tenets of operationalism, but merely to tame them.

5   The cleaning and washing of the IPK and the case for metrological operationalism

In preparation for each of the periodic verifications, the kilogram prototypes, including the IPK itself, were cleaned by a procedure documented by the BIPM (Thiesen, 1898; CIPM, 1946; Girard, 1990). For the third periodic verification, each prototype was first rubbed with a chamois leather cloth, which had been soaked three times, each for 48 hours, in a mixture of equal parts ethanol and ether, before the solvent was wrung out. A fairly hard pressure, estimated to be in the region of 10^4 Pa, was applied during the rubbing. Next, the prototype was steam washed to remove all traces of the solvent. For this, a jet of steam was sprayed directly at the surface of the prototype from an orifice of diameter 2 mm and at a distance of approximately 5 mm away. Any remaining water droplets were removed using an edge of filter paper.

Although a similar procedure had been performed before the first and second verifications, the third verification included an investigation, also conducted by Girard, into the effects of cleaning with solvent and steam washing upon the prototypes. Girard's results are shown in Figure 7.4, revealing that, in the months after being cleaned and washed, the mass of the platinum-iridium prototypes increased by 1 μg per month. As a result of this work, it was appreciated that a clarification to the 1889 definition of the kilogram was required. In 1989, the CIPM confirmed that the original definition of the kilogram referred to the IPK just after washing and cleaning by the official BIPM procedure; any comparison

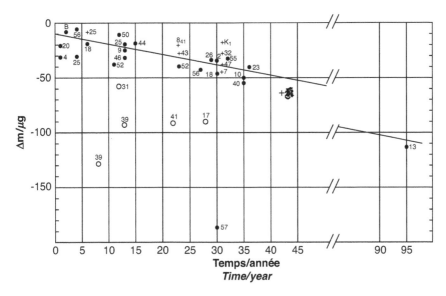

Figure 7.4 The change in mass of the kilogram prototypes on cleaning and washing, plotted against the years elapsed since their last cleaning and washing. Crosses identify the IPK and its six official copies. Open circles indicate prototypes with poor surface condition (Girard, 1990; CIPM, 1989, p. 130).

made to the IPK would therefore have to include an extrapolation to this mass. When accepting this proposal, the committee made it clear that this was not be interpreted as an alteration to the definition of the kilogram standard itself: "After considerable discussion, the Comité adopted this interpretation for the purposes of the third verification but made it clear that this did not in any way constitute a redefinition of the kilogram" (1989, p. 104). It did, however, enter the BIPM's *SI Brochure*, included in the instructions for realizing a kilogram (2006, App. 2).

The clarification highlights the current impossibility of knowingly obtaining a complete pointer (or *mise en pratique*) for a scientific standard. The metrological community is aware that a pointer is not entirely given by a short description and is willing to assume that an additional interpretation is required. In this case, a clarification that could have been added to the definition of the kilogram was instead included in its *mise en pratique*. The history of metrology is littered with occasions when a definition was instead altered in order to clarify its determination.[11] The logical empiricists were right, then, that the distinction between a pointer and its *mise en pratique* is not as clear as it is presented by the conventionalist account of scientific standards. Historically, the decision to choose a *mise en pratique* is bound up with the choice of a pointer, and it is not obvious where the line between the two should be drawn, if it is to have epistemic significance.

This was appreciated by generic conventionalists, and thus I conclude, with Reichenbach amongst others, that metrology takes its place in necessarily setting the framework for physical theory. Yet, as I have argued in the last section, generic conventionalism does not follow through the consequences of this point far enough. It is not just that metrology gives physical theory a conventional nature, but also the reverse: in addition, it performs an empirical role for the physical sciences. Echoing the point of Section 3, the selection of a pointer and its realizations is an investigation into the nature of reality.

The point can be considered from a semantic angle: is it the definition of a scientific standard or the procedure to realize the definition that gives meaning to that standard? The logical empiricists emphasized the importance of physical procedures in conferring meaning upon words. More generally, twentieth-century philosophers have suggested that meaning is to be found in the way a word is used, or the way a statement is verified, or the actions that are performed when applying a word.[12] Even for the most practical philosophers, however, this doctrine has not been brought to bear in a way that hinges the meaning of a scientific concept on the full and intricate network of metrological procedures by which we measure quantities of that concept. At most, a verificationist goes as far as finding the meaning of "It has a mass of so many kilograms", in the single, immediate procedure that a scientist would perform to check that result. In retrospect, an alternative is available to the philosopher seeking meaning in actions: the entire chain of calibrations, performed previously (and in most cases by other people), leading all the way back to the IPK, is where lies the meaning of "It has a mass of so many kilograms", and thus the significance of "kilogram" and then, perhaps, even "mass".

It is perhaps Percy Bridgman's writings, especially in his later work, that are most suggestive of this theory of meaning. He writes that, "the meanings of one's terms are to be found by an analysis of the operations which one performs in applying the term in concrete situations *or* verifying the truth of statements *or* in finding the answers to questions" (1938, pp. 114–131, emphasis added). Bridgman thus allowed for many different kinds of operations (including those he described as "mental" and "of paper and pencil") to confer meaning, although he never explicitly offered metrological procedures to fulfil this role. In the case of mass, for example, Bridgman assumed that the concept was best understood by contemplating procedures to measure force in the absence of a gravitational field – his suggestion was the deformation of elastic materials – and then extricating the concept of mass from this (1927, pp. 102–108). The proposal reveals a theoretical bias in even the most pragmatic of physicists. Why turn to Newton's equivalence between mass and force and not instead take the operations that are actually used in scientific practice to measure mass, which in Bridgman's time, as well as now, ultimately required the IPK, sitting upon a balance tray?

Bridgman did not intend to advocate any theory of meaning. A Nobel Prize winning physicist himself, he promoted his "operational attitude" as a way for colleagues to think more clearly about the concepts they dealt with. The kilogram never appeared on his radar. After all, it was obviously a term that lacked any conceptual confusion. Nor have scientific units, their manifestly conventional definitions being clearly stated and firmly agreed upon by the scientific community, appeared to philosophers to hold much philosophical promise.[13] Even the most anti-realist of thinkers have assumed a hierarchy between "mass" and "kilogram" (and thus accepted, unwittingly or not, a little realism): it is only the concept of mass that is troubled, once we define and understand what that is, whatever that may be, a kilogram is nothing more than a particular amount of it. Thus they have accepted the hierarchy of the conventionalist account of scientific standards: a definition of a unit is relegated below that of the quantity it measures. Anti-realism is thus forced to take on a two-sided conventionalism that rather mimics the realist position it was intended to oppose. Although the verificationist or operationalist looks to procedures, actions or operations to reveal what mass is, these are separated from those that reveal what a kilogram is, the former being required prior to and independently of the latter. When the scientist, in a particular context, turns only to a particular metrological procedure to take a mass measurement, the verificationist looks to that procedure to provide only the meaning of "kilogram", and then looks through the textbooks to discover what could possibly be used to provide a further meaning for "mass". On reflection, this is a peculiar task because the metrological work that generates the kilogram also gives science its processes for measuring mass. Furthermore, in practice, we do not need to be committed to a particular view about mass, but only to make assumptions about what kinds of thing retain their mass over time, in order to define a mass unit. In practice, procedures to make measurements – at least preliminary ones – come before a theoretical understanding of what is being measured.

The operational attitude of Bridgman – as well as the like-minded anti-realism of the generic conventionalist – thus faces the problem that, despite first appearances, it has not achieved what it set out to do in overcoming the crude distinction of naïve realism between the objective and the subjective. It upholds the realist's hierarchy between a quality and its standard, using different operations to determine the meaning of each. To retain an internal simplicity and coherence to the thesis of operationalism – to go the whole way – it is necessary to look to all the metrological procedures associated with a standard to give meaning, not just to that standard, but the quality it represents. This alternative, more thoroughgoing operationalism reverses the dependence between "kilogram" and "mass". It overcomes one of the main difficulties facing Bridgman's theory, by clarifying what counts as an "operation" (which is now understood as a procedure used by the scientific community, ratified by metrologists, to make measurements). In embracing all metrological procedures as holding significance, thoroughgoing operationalism makes little distinction between calibrations and the precision measurements of experiment. What is more, by more vigorously shaking off conventionalism regarding scientific standards, it provides a stronger base from which to respond to the problem of accounting for metrological progress, considered so far with respect to the scientific-standard conventionalist. The problem presents itself even more starkly against the operationalism of Bridgman. Donald Gillies (1972, pp. 6–7) raised the point most clearly: if we declare the meaning of scientific measures to be found in the bare metrological procedures for their realization, why would we ever want to improve a method of measurement?

A logical empiricist, verificationist, operationalist or even a more thoroughgoing operationalist – appears to be truly stuck. In my interpretation of the issue in Section 3, however, I have argued that the naïve realism associated with the conventionalist account of scientific standards faces the same issue. The fix is not to add a dose of realism about scientific posits alongside a conventional view of scientific standards, attempting to explain metrological progress as the move towards a truer or more objective measure. Metrologists are not redefining the kilogram after having been presented with a more accurate account of what mass is. It was their decision to reject alternative definitions of the kilogram. The "thick" motivation of stability – the fact that the better definition is revealed by improved metrological precision – is not accounted for by half-half vision of science (what it measures is real; how it measures it is convention). Gillies' objection does not only apply to operationalism; neither does a simple realistic picture of science accurately portray the progress of metrology. What is more, I believe that the reason Gillies' argument is so arresting is that it supposes that operationalism adheres to certain elements of the conventionalist account of scientific standards. The operationalist is assumed to be a perfect conventionalist in that the operations chosen to provide meaning must necessarily be chosen for their practical advantages only. When we realize that these operations are providing meaning precisely because the assumptions implicit in those operations – the conservation of mass of an iridio-platinum piece, for example – are true to reality, we can see that an operationalist reforms a scientific standard to improve the stability of measurement, a notion that is at once pragmatic and realist.

Thoroughgoing operationalism is far too cumbersome as it stands, however. I do not advocate it in its entirety but tame it as much as I can: *metrological operationalism* is the thesis that the meaning of a scientific concept is at least partly given by the collection of all methods by which it is realized, including the *mise en pratique* for its scientific standard. This is the semantic version of the thesis that metrology contributes empirical content to science; I propose it as a possible contribution to a metrological account of scientific standards.

6   A metrological account of scientific standards

The discussion of this paper is intended to motivate the larger question: what role do scientific standards play in the development of physical theory? I take a metrological account of scientific standards to be a response to this that takes into consideration the practice, both current and historical, of metrology. I have argued that this is necessary because the conventionalist account of scientific standards, most clearly associated with realist thinking but also identifiable in anti-realist thought and reaching further into our philosophical thought than is immediately apparent, fails to account for metrological progress. I have made tentative suggestions for a metrological account of scientific standards. It might reasonably include a version of metrological operationalism (but would in any case acknowledge that the meaning of a standard is not wholly contained in its official definition). It might reasonably support metrological realism regarding scientific standards (but would in any case acknowledge a deeper interplay between metrology and physical theory than conventionalist and realist thinking). It might reasonably view the physical sciences to be unified by metrology (but would in any case acknowledge that metrology sets the empirical limits of the physical sciences or even contributes to its empirical content).

The overlaying of the history of metrology upon that of twentieth-century analytic philosophy that I have applied here is not merely a series of analogies (between the results of calibration and analytic statements, between the results of experiment and synthetic statements, between the procedure of translation and those of calibration). Metrology offers a testing ground for theories of meaning. The conventionalist account of scientific standards has stood in the way of undertaking real metrological examples to philosophy of science and philosophy more generally. Elements of this account resonate in philosophical thought, despite the fact that it is at odds with more general conventionalist thinking. Reichenbach was perfectly right to point out particular conventions are necessary to get measurement off the ground and ultimately sat disguised within the body of physical theory. It is not an observation that generalizes well, however: the resulting conventionalism ultimately had the unfortunate side-effect of sweeping metrology beneath the carpet. The search for convention entering into theory and experiment should be complemented with a search for the theoretical and empirical resulting from measurement. (The point, after all, has always been that "theoretical" and "conventional" are not terms that stick fast in philosophy.) Here, I have attempted to work towards that goal, in a way that echoes Peter Galison's focus on an experiment's end.

Despite the lack of a deductive closure to experiment, Galison argued that rising experimental evidence eventually becomes sufficiently persuasive to draw an experiment to an end (1987). He argued that the complicated tangle of factors that brought experiments to this point had to be unravelled from a historical perspective, discouraging simplistic philosophical models. Mimicking the logician's crude closure of an experiment, the scientific-standard conventionalist assumes simple distinctions (between an experiment and its associated calibrations, between the empirical nature of physical theory and the conventional nature of metrology, between the accuracy of the idealist and the precision of the pragmatist) to mark the beginning of an experiment. It is only after we know what we are measuring and how we are measuring it, that experimental science may begin. I've been arguing here, then, that a more detailed, historical approach is also required in order to untangle the factors that bring experiments to their beginning.

Acknowledgements

I would like to thank two anonymous reviewers for their particularly insightful comments. I would also like to thank Jeremy Butterfield, Hasok Chang, Jim Grozier, Eran Tal as well as all other members of the Philosophy and History of Science reading group at the University of Cambridge. I thank David Lloyd for the creation of Figure 7.3.

Notes

1 The motivations for the creation of the original kilogram are explored by Riordan (2015).
2 The motivations for the creation of the original metre are explored by Kershaw (2012).
3 The state of measurement in England at this time is surveyed by Schaffer (1995).
4 No. 25 and No. 31 have since been renumbered.
5 This hypothesis has been articulated by Terry Quinn, director of the BIPM between 1988 and 2003 (2011, p. 365).
6 Natural philosophers who supported using more fundamental definitions of scientific standards include Picard (1671), Huygens (1673), Wilkins (1688), de La Condamine (1747), Whitehurst (1787) and Lavoisier (1893). The metric project was initiated by a call for natural scientific standards, recorded by the editors of the parliamentary archives, Mavidal and Laurent (1881, pp. 104–108). The resolutions of the CGPM show the desire for "a natural and indestructible standard" (1961, p. 85), "a natural base" (1949, p. 44) and to define scientific standards "in terms of the invariants of nature" (2010, p. 434). The desire is also reflected in articles of modern metrology, including Blevin and Steiner (1975), Kibble and Robinson (2003) and Mills et al. (2011).
7 A survey of the changing meaning of a fundamental standard is given by Riordan (2015).
8 Joseph Mulligan gives a brief account of the lives and work of Pérot and Fabry (1998).
9 De Jacobi, member of the Imperial Academy of Science, Saint Petersburg, brought up the possibility of using a collection of artefacts to define the kilogram at a meeting of the International Metre Commission in 1872 (Quinn, 2011, p. 52).
10 I am appropriating the notion of a thick concept used in the ethics of science, in which a factual element as well as one of value can be found, a discussion that goes back to Ernest Nagel (1979, pp. 485–502).
11 Examples include the 1927 amendment to the metre described in Section 1.
12 There are many philosophical works that could be used to support my claim here. I am thinking in particular of those of A. J. Ayer (1936), P. W. Bridgman (1938), Ludwig Wittgenstein (1953), W. V. O. Quine (1960) and Michael Dummett (1978).
13 A notable exception is Wittgenstein's insistence that it cannot be said of a metre stick that defines the metre that, "it is one metre long" (1953, §50), and Kripke's ensuing suggestion that such statements are *a priori* contingent truths (1980).

References

Avetissian, A. K. (2009). Planck's constant variation as a cosmological evolution test for the early universe. *Gravitation and Cosmology*, *15*, 10–12.

Ayer, A. J. (1936). *Language, truth and logic*. London: V. Gollancz, Ltd.

Bell, S. (2001). *A beginner's guide to uncertainty of measurement: Measurement good practice guide, No. 11, Issue 2*. Teddington, Middlesex: National Physical Laboratory.

BIPM. (2006). *SI Brochure: The International System of Units (SI)* (8th ed.), updated in 2014. Paris: Bureau International des Poids et Mesures. Retrieved from www.bipm.org/en/publications/si-brochure.

BIPM. (2008). *Evaluation of measurement data: Guide to the expression of uncertainty in measurement*. Joint Committee for Guides in Metrology, 100. https://www.bipm.org/utils/common/documents/jcgm/JCGM_100_2008_E.pdf.

BIPM. (2012). *International vocabulary of metrology: Basic and general concepts and associated terms (VIM)*. Joint Committee for Guides in Metrology, 200. https://www.bipm.org/utils/common/documents/jcgm/JCGM_200_2012.pdf

BIPM. (2013). *Draft Chapters 1, 2 and 3 of the 9th SI Brochure*. Paris: Bureau International des Poids et Mesures.

Blevin, W. R., & Steiner, B. (1975). Redefinition of the Candela and the Lumen. *Metrologia* 11, 97–104. http://doi.org/10.1088/0026-1394/11/3/001.

Bridgman, P. W. (1927). *The logic of modern physics*. New York, NY: Macmillan.

Bridgman, P. W. (1938). Operational analysis. *Philosophy of Science*, 5, 114–131.

Carnap, R. (1950). Empiricism, semantics, and ontology. *Revue Internationale de Philosophie*, 4, 20–40.

CGPM. (1890). *Comptes rendus des séances de la première conférence générale des poids et mesures, réunie à Paris en 1889*. Paris: Gauthier-Villars.

CGPM. (1928). *Comptes rendus des séances de la septième conférence générale des poids et mesures, réunie à Paris en 1927*. Paris: Gauthier-Villars.

CGPM. (1949). *Comptes rendus des séances de la neuvième conférence générale des poids et mesures, réunie à Paris en 1948*. Paris: Gauthier-Villars.

CGPM. (1961). *Comptes Rendus de la 11e conférence générale des poids et mesures, réunie à Paris en 1960*. Paris: Gauthier-Villars.

CGPM. (1975). *Comptes rendus des séances de la quinzième conférence générale des poids et mesures, Centenaire de la Convention du Mètre et du Bureau International des Poids et Mesures, Paris 27 Mai – 2 Juin 1975*. Paris: Bureau International des Poids et Mesures.

CGPM. (2010). *Comptes rendus de la 23e réunion de la Conférence générale des poids et measures (novembre 2007)*. Paris: Bureau International des Poids et Mesures.

Chang, H. (2004). *Inventing temperature: Measurement and scientific progress*. Oxford Studies in the Philosophy of Science. New York, NY: Oxford University Press.

CIPM. (1946). *Procès-verbaux*, Series 2, Vol. 20. Comité international des poids et mesures.

CIPM. (1989). *Procès-verbaux de la 78e session*, Vol. 59. Comité international des poids et mesures.

CIPM. (1993). *Procès-verbaux de la 82e session*, Vol. 61. Comité international des poids et mesures.

CIPM. (2008). *Procès-verbaux de la 97e session*. Comité international des poids et mesures.

Davies, R. S. (2005). Possible new definitions of the kilogram. *Philosophical Transactions of the Royal Society A: Mathematical, Physical and Engineering Sciences*, 363, 2249–2264.

de La Condamine, C. M. (1747). *Nouveau projet d'une mesure invariable propre à servir de mesure commune à toutes les nations*. Paris: Académie royale des sciences.

Dirac, P. A. M. (1937). The cosmological constants. *Nature*, 139, 323.

Dummett, M. (1978). *Truth and other enigmas*. London: Duckworth.

Eichenberger, A., Geneves, G., & Gournay, P. (2009). Determination of the Planck constant by means of a watt balance. *The European Physical Journal Special Topics*, 172, 363–383.

Galison, P. (1987). *How experiments end*. Chicago, IL: University of Chicago Press.

Gill, D. (1907). Address of the President of the British Association for the Advancement of Science. *Science*, New Series, 26(659), 193–212.

Gillies, D. (1972). Operationalism. *Synthese*, 25, 1–24.

Girard, G. (1990). *The procedure for cleaning and washing platinum-iridium kilogram prototypes used at the Bureau International des Poids et Mesures*. Paris: Bureau International des Poids et Mesures.

Hill, T. P., Miller, J., & Censullo, A. C. (2011). Towards a better definition of the kilogram, *Metrologia*, *48*, 83–86. http://doi.org/10.1088/0026-1394/48/3/002.

Huygens, C. (1673). *Horologium oscillatorium: Sive, de motu pendulorum ad horologia aptato demostrationes geometricae*. Paris: F Muguet.

Kelley, M. H. (2001). *The electronic kilogram*. Microwave Symposium Digest, IEEE MTT-S International, pp. 857–860. https://www.nist.gov/publications/electronic-kilogram

Kershaw, M. (2012). The 'nec plus ultra' of precision measurement: Geodesy and the forgotten purpose of the Metre Convention. *Studies in History and Philosophy of Science Part A*, *43*, 563–576.

Kibble, B. P. (1976). A measurement of the gyromagnetic ratio of the proton by strong field method. In J. H. Sanders & A. H. Wapstra (Eds.), *Atomic masses and fundamental constants* (Vol. 5, pp. 545–551). New York, NY: Plenum Press.

Kibble, B. P., & Robinson, I. (2003). Replacing the kilogram. *Measurement Science and Technology*, *14*, 1243–1248.

Kripke, S. (1980). *Naming and necessity*. Cambridge, MA: Harvard University Press.

Lavoisier, A. L. (1893). Eclaircissements historiques sur les mesures des anciens. In M. Dumas & E. Grimaux (Eds.), *Oeuvres de Lavoisier, 1864–1893* (Vol. 6, pp. 698–705). Paris: Imprimerie Nationale.

Mavidal, J., & Laurent, E. (1881). *Archives parlementaires de 1789 à 1860; recueil complet des débats législatifs & politiques des chambres française*. Series I (Vol. 12). Paris: Dupont.

Maxwell, J. C. (1870). *Address to the Mathematical and Physical Sections of the British Association*. Report of the Fortieth Meeting of the British Association for the Advancement of Science, 40, Notices and Abstracts, pp. 1–8.

Mills, I. M., Mohr, P. J., Quinn, T. J., Taylor, B. N., & Williams, E. R. (2011). Adapting the International System of Units to the twenty-first century. *Philosophical Transactions of the Royal Society A: Mathematical, Physical and Engineering Sciences*, *369*, 3907–3924.

Mulligan, J. (1998). Who were Pérot and Fabry? *American Journal of Physics*, *66*, 9.

Nagel, E. (1979). *The structure of science*. Indianapolis, IN: Hackett.

Picard, J. F. (1671). *Mesure de la terre*. Paris: Imprimerie Royale.

Quine, W. V. O. (1960). *Word and object*. Cambridge, MA: MIT Press.

Quinn, T. (2011). *From artefacts to atoms: The BIPM and the search for ultimate measurement standards*. Oxford: Oxford University Press.

Reichenbach, H. (1928). *Philosophie der Raum-Zeit-Lehre*. English translation by Maria Reichenbach (1957): *The philosophy of space and time*. New York, NY: Dover Publication.

Riordan, S. (2015). The objectivity of scientific measures. *Studies in History and Philosophy of Science*, *50*, 38–47.

Schaffer, S. (1995). Accurate measurement is an English science. In N. Wise (Ed.), *Values of precision* (pp. 135–172). Princeton, NJ: Princeton University Press.

Steiner R. L., Williams E. R., Newell D. B. and Liu R. (2005) Towards an electronic kilogram: An improved measurement of the Planck constant and electron mass. *Metrologia*, *42*, 431–41.

Thiesen, M. (1898). *Travaux et mémoires du Bureau international des poids et mesures*, 9.

Whitehurst, J. (1787). *An attempt towards obtaining invariable measures of length, capacity, and weight, from the mensuration of time, independent of the mechanical operations requisite to ascertain the center of oscillation, or the true length of pendulums.* London: W. Bent.

Wilkins, J. (1688). Of measure. In *An essay toward a real character and a philosophical language*. London: S. Gellibrand.

Wittgenstein, L. (1953). *Philosophical investigations* (G. E. M. Anscombe & R. Rhees, Ed. and G. E. M. Anscombe, Trans.). Oxford: Blackwell.

8 The SI and the problem of spatiotemporal constancy

Ingvar Johansson

1 The problem of spatiotemporal constancy

Crucial to the establishment of good measurement units is to find or construct something that can be regarded as being constant. What is one metre ought to be one metre everywhere and at all times. Metre sticks ought neither to expand nor contract; and if they do, there should somewhere in a calibration hierarchy be a prototype or a naturally given length standard that – presumably – never changes length; and against which the changing sticks now and then can be re-calibrated.

I will call the search for constancies that can ground fundamental measurement units *the problem of spatiotemporal constancy*, and I find it a bit remarkable what small attention the problem has received in the philosophy of science. The explanation, I guess, is that the problem of such constancy is very much a substantial problem, and that philosophers of science have mainly rested content with discussing formal-logical and set-theoretical features of measurement procedures.[1] This chapter tries to remedy this fact.[2] It is concerned with how the problem of constancy is, and has been, handled in the International System of Units, henceforth the SI.

The present SI (SI8, 2006) divides its standard units for measurement into base units and derived units, a distinction that I will in detail discuss.[3]

2 The New SI

A draft of the second chapter of the so-called New SI was signed in September 2010 (SI9 Draft, 2010), a specification was put forward in March 2013, in December 2013 a preliminary version of the first three chapters appeared (SI9 Draft, 2013) and in November 2016 the last version that I have had recourse to appeared (SI9 Draft, 2016). A decision is expected by 16 November 2018. The proposal did by no means reach immediate acclamation; rather, the contrary, as witnessed by the website *Metrology Bytes* (2012). There exists even today among metrologists a whole spectrum of views towards the New SI.

At the positive end of the spectrum we find, of course, the proposal authors. They have talked about the "extraordinary advances [that since 1960] have been made in relating SI units to truly invariant quantities such as the fundamental constants of physics and the properties of atoms" (SI9 Draft, 2013, p. 8); they look upon their proposal as a culmination of this trend. The New SI wants to take away the last existing material prototype, the international prototype of the kilogram. Instead, according to the 2013 version, they want to relate all the base units to constants in fundamental laws and to properties that are invariant when derived from fundamental laws (SI9 Draft, 2013, p. 11). However, in the 2016 version they use the generic term "defining constants" (SI9 Draft, 2016, p. 2), and these constants are then claimed to be of four different kinds: (a) fundamental constants of nature, (b) specified atomic parameters, (c) conversion factors and (d) technical constants (SI9 Draft, 2016, p. 4).

From the negative end of the spectrum I will quote a chemist and metrologist who labels himself *advocatus diaboli* (Price, 2010, p. 421; 2011, p. 131a). He writes: "The choice of the fundamental constants of nature as metrological anchors, as they are understood by current science, at current best accuracy values, runs a real risk of being a Zanzibar system of cosmic proportions" (Price, 2011, p. 130a). A Zanzibar system where, *unknown to each other*, A checks his measuring device (in the original story a clock) by means of the device belonging to B, who, conversely, checks his instrument against A's; described by e.g. Crease (2011, intr.). If B's device starts to malfunction after he has checked it against A's, but before A calibrates his against B's, the malfunction might not be detected.

In the Zanzibar system the measurement instruments are of the same kind (clocks), but in relation to the New SI the fear is that the standard units for different kinds of quantity have become so interdependent that their spatiotemporal constancy can no longer in any real sense be tested. The indisputable point is that the links the New SI creates between base units and defining constants, praised by the defenders of the New SI, bring with them several new interdependencies between the old units. This paper is an attempt to sort out and clearly distinguish the different kinds of circularities and dependencies that exist in both the older SI systems and the New SI. Let it be said at once, there is a semantic issue that I find to be beyond all doubt: it has become directly misleading to talk about a distinction between two kinds of standard units, base units and derived units, respectively (see Section 7).

In this chapter, I will in detail comment upon the new definition proposals for the metre, the ampere and the kilogram.[4]

3   Inevitable semantic circularities

Standard units, be they called base units or derived units, are always units for what the SI calls *quantities*, but what for the pedagogical purposes of this chapter I will from now on (quotations aside) call *kinds-of-quantity*.[5] For instance, the second is the unit for the kind-of-quantity time duration, the metre is the unit for length, the kilogram is the unit for mass and the metres-per-second is the unit for velocity. All base unit definitions take a kind-of-quantity for given (SI8, 2006, p. 103). Therefore, the first thing to be noted in relation to possible circularities is that no concept of a kind-of-quantity can be understood before it is connected to other concepts.

It is always the case that in order to understand a certain concept we also have to understand a bunch of other related concepts. This non-atomistic view of concept meaning is part and parcel of modern philosophy of language. It is even a view that crosses some traditional philosophical divisions. Within analytic philosophy, it can very explicitly be found in philosophers as different as Ludwig Wittgenstein, W.V.O. Quine and Donald Davidson; and within non-analytic philosophy a semantic non-atomism is fundamental to both hermeneutic philosophy and structuralism. It permeates modern cognitive science, too.

Trivially, we cannot understand "to the left" without understanding "to the right", and vice versa. We have to move in a circle and learn both the concepts simultaneously. Similarly, but a bit more complex, no one can learn what "mass" means in Newtonian mechanics without at the same time learning about some other concepts in Newtonian mechanics, too. In order to understand what "mass" refers to, we must understand at least the referential difference between the concepts "mass" and "weight", and in order to understand this difference, we have to learn a little also about the concept "gravitational force" – weight being equal to mass times gravitational force. The concept "mass" cannot be introduced by means of a definition where all the defining terms are already understood, old, everyday laymen concepts.

Here is another simple example. For a long time, the basic unit for measuring temporal duration was the solar day. But the concept of solar day relies on concepts of spatial distances. A solar day is the time interval between two successive appearances of the Sun at its *highest point* in the sky. In a similar way, anyone trying to understand a concept that refers to one of the present base kinds-of-quantity of the SI will have to learn a network of concepts.

No definition of a standard unit can possibly be regarded as simultaneously supplying a definition of the kind-of-quantity that it is construed to be the standard unit for. "The unit is simply a particular example of the quantity concerned" (SI8, 2006, p. 103); "For a particular quantity different units may be used" (SI9 Draft, 2016, p. 2). A standard unit can be defined only when there is a pre-given understanding of a concept that refers to a kind-of-quantity. And, always, this concept is part of some conceptual networks where loops are the rule rather than the exception.

I will call this fact *the semantic-holistic predicament of human measurement.* Necessarily, all editions of the *SI Brochure* have conformed to it, and all future editions will.

4   Inevitable epistemic circularities

During the twentieth century, not only the philosophy of language moved from atomistic to non-atomistic views; the same happened with the view on empirical testing in the philosophy of science. Karl Popper stressed the need for background knowledge, Thomas Kuhn coined the term "paradigm" and Imre Lakatos the term "research program". This development took place in parallel with a growing general acceptance of Quine's expression of "the underdetermination of scientific theories by empirical evidence". This means that even the presumed constancies of c (the scalar velocity of light in vacuum) and h (the Planck constant) are underdetermined by empirical evidence. The first constant is part and parcel of the relativity theories, and the second of the quantum-mechanical theories. Empirical science is fallible through and through; the presumed constancies of c and h are no exceptions. But the same is equally true of the constancy of metrological prototypes. Let me explain.

The present SI definition of the kilogram can be given this simple verbal form:

> 1 kilogram mass $=_{\text{def}}$ the mass of the platinum-iridium cylinder at BIPM (and then, by implication, all other bodies that have the same or an exactly similar mass have a mass of 1 kg, too).[6]

As pointed out in Section 3, in order to understand this definition we have to understand what mass is. But there is more complexity to this seemingly straightforward definition than the conceptual circles that come with the mere concept of mass. The purely verbal definition does not mention why the cylinder is made by platinum-iridium, why its edges have the form they have, why there are handling instructions and why there are six official copies of the prototype against which the prototype can be compared. The kind of *surround knowledge* (if I may coin a term) that is used in order to secure that the prototype does not change mass over time, is not mentioned in the explicit verbal definition, but without such knowledge the definition would be completely gratuitous, and thereby useless. A reasonable prototype definition of the kilogram must implicitly rely on knowledge of other kinds-of-quantity than that of mass; in particular, on some law-like connections between these and possible changes of mass.

No construction and definition of an SI prototype can possibly avoid the kind of knowledge dependence between kinds-of-quantity now noted. All material prototypes are in principle vulnerable to deterioration or damage. Now, of course, the fundamental constants c and h are not vulnerable to such changes; the uncertainty of their constancy comes, as said, from the empirical underdetermination of scientific theories. However, not even on the assumption that c and h will forever be part of our best tested theories, is there an epistemic gulf between *constant based* unit definitions and *prototype based* unit definitions. A constant based definition would be completely useless if there are no scientists who know how to put the definition into experimental practice. Each constant based definition must be, and is, complemented by a so-called *mise en pratique*, i.e., a set of instructions of how to create a primary realization of the definition in question (the French expression is used also in English SI texts). The knowledge contained in such instructions can be called a kind of surround knowledge, too.

In the present SI, the *mises en pratique* are presented in an appendix that is available only online (SI8, 2006, p. 172), but they will be given a more central position if the New SI (or something like it) becomes accepted. The 2013 version of the proposal says rightly that it "effectively decouples the definition and practical realization of the units" (SI9 Draft, 2013, p. 9), which implies that it cannot rest content with talking only about definitions.[7]

Put briefly, *what the handling instructions are to a prototype based standard unit, the primary* mises en pratique *are to a constant based standard unit.* Measurement units cannot possibly be wholly anchored in a theoretical realm; not even if the empirical underdetermination of theories is neglected. Surround knowledge of the *mise en pratique* kind will still be needed.

What I have now described might be called *the double epistemic-fallibilistic predicament of human measurement.* (The term "double" indicates, on the one hand, a reference to the fallibility of the testing of the theories behind constant-based unit definitions and the construction knowledge behind the prototypes, and, on the other hand, a reference to the fallibility of the *mises en pratique* of constant-based units and the handling instructions of the prototypes.) Necessarily, all editions of the *SI Brochure* have conformed to this predicament, and all future editions will.[8]

5 Contingent dependencies between base units

The *inevitable* semantic and epistemic dependencies highlighted in the last two sections should be kept distinct from the kind of *contingent* dependence I will mention in this section, namely the kind of dependence that exists when a base unit in its definition explicitly or implicitly mentions some other base unit(s). Definitions of *derived* units must of course in some way mention the base units from which they are derived, but derived units will be discussed in the next section.

The kind of dependence now at issue has a long history. For instance, when in 1795 in France the gram was officially defined, it was decreed to be the weight of a volume of one cubic *centimetre* of pure water (at the temperature of melting ice).[9] In the present SI it looks as follows: (a) the metre definition is dependent on that of the second, but not vice versa; (b) the ampere definition is dependent on those of the second, the metre and the kilogram, but the latter three are not dependent on the ampere; and (c) the mole definition is dependent on the kilogram, but not vice versa. This is explicitly stated:

> Finally, it should be recognized that although the seven base quantities – length, mass, time, electric current, thermodynamic temperature, amount of substance, and luminous intensity – are *by convention* regarded as independent, their respective base units – the metre, kilogram, second, ampere, kelvin, mole, and candela – are in a number of instances interdependent.
>
> (SI8, 2006, p. 111, emphasis added)

However, I find the term "interdependent" too wide. I have inserted the expression "but not vice versa" in order to make it clear that some of the dependencies referred to are *unilateral*. In the next section, on the other hand, we will also meet *mutual* dependence relations between standard units.

In the New SI, leaving the dependencies on fundamental constants aside, it looks as follows: the metre definition is dependent on that of the second, but not vice versa; the ampere definition is dependent on that of the second, but not vice versa; the kilogram is dependent on the second and the metre, but not vice versa. That is, even the New SI contains some unilateral base-to-base unit dependencies.

This kind of base-to-base unit dependency is not inevitable. In principle, it would be possible to make one prototype for each base kind-of-quantity, but the existence of the inevitable epistemic circularities mentioned makes it a matter of pragmatics whether base-to-base unit dependencies should be allowed or not. I can at the moment see no general reason to try to get rid of them only because they are not inevitable. They have to be discussed on a case by case reasoning, and the same is true of the dependencies discussed in the next section.

6   Base units as dependent on a constancy in a derived kind-of-quantity

When prototypes such as the classical standard metre and standard kilogram are used in metrology, it becomes natural and practical to let the assumedly unchanging prototype be a direct exemplification of the standard unit, and so ascribed the number 1. But there is no theoretical necessity behind this choice. When a unit grounding constancy is found in nature, it is often, to the contrary, natural and practical to make the standard unit a *fraction* or a *multiple* of the assumed constancy. For instance, the French Academy of Sciences declared in 1791 the metre to be the *fraction* one tenth-millionth of the assumedly constant length of the meridian going from the North Pole through Paris to the Equator; and between 1960 (when the metre prototype was dropped) and 1983 the SI defined 1 metre to be equal to the *multiple* 1 650 763.73 wavelengths of an assumedly constant radiation (that corresponding to the transition in a vacuum between the $2p^{10}$ and $5d^5$ quantum levels of the krypton-86 atom). Base units are intended to be grounded in a magnitude that is spatiotemporally constant, but such a constancy needs not to be ascribed the number 1 and regarded as a direct exemplification of the unit it grounds.[10]

What has just been highlighted is seldom explicitly stated in metrological writings, but it is important to keep it in mind in what follows. In three different cases with one subsection for each, dependencies between definitions of base units and derived units in the SI will be discussed.

A base unit definition is neither a definition that relates a concept to other concepts as in a *lexical definition*, nor a so-called *Aristotelian real definition*; the latter lays claim to define what the true nature of a certain kind of objects is like. A base unit definition is a *coordinative definition* (Reichenbach, 1958, §4); it stipulates a relation between a concept on the one hand and a magnitude, object or kind of objects in the language-external world on the other.

Since the concepts used (e.g., "1 kg") always contain a number reference, the definitions connect numbers with the world, too. All the base unit definitions of the New SI start with a sentence that conforms to the following abstract form:

"The base unit U, symbol u, is the SI unit of kind-of-quantity *Q.* It is defined by
. . .". Talking in terms of "*the* metre", "*the* kilogram", etc. ("*the* base unit U"),
instead of "1 metre", "1 kilogram", etc., hides the fact that the number 1 is part
of the definition; but, surely, the number is implicitly there.

6.1   The case of the metre and the metres-per-second

In both all the older SI systems and in the New SI, it is taken for granted that the
relationship $v = l/t$ obtains. If standard units are stipulated for two of these three
kinds-of-quantity (v, l, t), then the third kind-of-quantity gets a standard unit so
to speak for free. Let me in what follows presuppose that there is a good pre-given
definition of the standard unit for time duration, the second, and then consider
the different ways the metre has been defined.

After 1889 the metre has been given three different definitions. Between 1889
and 1960 it was defined as being equal to the length magnitude of the standard
metre in Paris, and between 1960 and 1983 it was defined as being equal to the
length magnitude of 1 650 763.73 wavelengths of a certain radiation. That is,
before 1983 the base units for l and t were stipulated, and the standard unit for
v was derived by means of the relationship $v = l/t$. I have no more to say about
these two definitions than what has been said in the sections about semantic and
epistemic circularities.

However, from a logical point of view, one can equally well define the standard
unit for v as being a certain fraction of the (according to relativity theory) constant velocity/speed[11] of light in vacuum, c and then let the standard unit of l,
the metre, be defined by means of the relationship $l = v\ t$. This is implicitly done
in the 1983 definition, and explicitly in the New SI. Ever since 1983 the metre
is grounded in the assumption that c is a natural constant. In the present SI, the
metre definition looks as follows:

> The metre [= 1 metre] is the length of the path travelled by light in vacuum
> during a time interval of 1/299 792 458 of a second.[12]

And then, in a new paragraph directly after the definition, it is stated:

> It *follows* that the speed of light in vacuum is exactly 299 792 458 metres
> per second.
>
> (SI8, 2006, p. 112, emphasis added)

From a logical-structural point of view, *the introduction of this metre definition
represents a break with older metrology, and it is this structure that is made pervasive
in the New SI.* Superficially, it may look as if this metre definition depends *only* on
the definition of the second, but this is wrong, which reflections on the last part
of the quotation can make clear. Nothing numerical about the velocity of light
can possibly *follow* from the metre definition if, beside time (t) and length (l), not
also velocity (v) is ascribed a standard unit. Surely, a simple constancy claim such

as the statement "the velocity of light is always and everywhere the same" needs no unit, but that is quite another thing. Obviously, in the metre definition a specific standard unit of velocity is tacitly presupposed, namely metres-per-second.[13] Moreover, it is also taken for granted that the velocity of this standard unit is 1/299 792 458 of the velocity of c.

In order to see the whole metrological structure of the metre definition, one must become clear about from where the metres-per-second comes, and why its sudden appearance looks so innocent.

It is all too easy to think that since $v = l/t$, then as soon as the metre is made the unit for length and the second for time, the unit for velocity cannot be but metres-per-second. The complete structure, however, looks as follows. The formula $v = l/t$ states a numerical relation, but it is not a purely mathematical relation. The variables are variables for physical magnitudes, and the formula cannot be applied to the world before the variables are connected to measurement units. Nonetheless the formula does not mention any specific measurement units; this is not a feature peculiar to this kinematic formula. The same goes for all the natural laws of mathematical physics; Newton's second law, $F = ma$, does not mention any measurement units either. We stumble upon a general problem that can be formulated thus: *how can there be numerical physical relationships without measurement units?* Let us take a closer look at $v = l/t$.

Implicitly, all the three variables are variables for ratios or proportions. The variable v represents the ratio between velocity magnitudes and *some* standard unit magnitude for velocity. Similarly, l represents the ratio between length magnitudes and *some* standard unit magnitude for length, and t represents the ratio between temporal durations and *some* standard unit magnitude for temporal duration. Or, to quote the SI: "[a] number [of a variable] is the ratio of the value of the quantity to the unit. For a particular quantity, many different units may be used" (SI8, 2006, p. 103).

Sometimes, kinds-of-quantity variables such as l, t and v are in metrology subsumed under a generic variable called Q, and what has just been said is stated by means of the following (non-arithmetic and non-algebraic formula): $Q = \{Q\}[Q]$ (SI9 Draft, 2016, p. 3).[14] Here, $[Q]$ symbolizes a unit for the kind-of-quantity Q, and $\{Q\}$ symbolizes the real numbers that correspond to the ratios between specific magnitudes and the unit $[Q]$. In other words, to claim that a certain magnitude equals the quantity Q_1 (e.g., l_1, t_1, or v_1) is to claim that it is Q_1 (e.g., l_1, t_1, or v_1) *times* a given specific unit magnitude.[15]

In this symbolism, the relation $v = l/t$ becomes $\{v\}[v] = \{l\}[l]/\{t\}[t]$. And by means of this formula a fact of importance can easily be made visible. In order for there to be equality between the left and the right hand sides, we cannot put in any units we want. If we let $[l]$ be the metre, $[t]$ be the second and $[v]$ be the yards-per-second, then $v = l/t$ is false. True is instead now: $v = 0.9144^{-1} l/t$ (i.e., v yards-per-second = 0.9144–1/metres/t seconds, since 1 metre equals 0.9144–1 yards).

A *completely unit independent* symbolization of the relation between (mean) velocity, distance traversed and the time of the movement cannot possibly be written $v = l/t$. However, two different kinds of changes can make the formula unit independent. Either the equality sign is exchanged for a proportionality sign: $v \propto l/t$, or a symbol for a purely metrological *unit adjuster* (such as 0.9144^{-1} in the example) is inserted in the formula. Such a number has the function retrospectively to secure that the equality holds whatever units are chosen for the variables. If we symbolize such a unit adjuster α, then we can in general truly claim:

$$v = \alpha \, l/t \text{ (which is shorthand for } \{v\}[v] = \alpha \, \{l\}[l]/\{t\}[t]).$$

The physical relation $v = \alpha \, l/t$ can be reduced to the equality $v = l/t$ only on the presuppositions that (a) the measurements units are chosen in such a way that $\alpha = 1$, and that (b) the number 1 is allowed to be absent from the formula. This kind of reduction has been both expedient and very fruitful in theoretical physics, but it may nonetheless invite bad metrological thinking.[16]

According to my experiences, it needs now and then to be stressed that more than pure mathematics is needed in order to go from the unavoidable view that mean velocity is directly proportional to distance traversed (given the time) and inversely proportional to the time needed (given the distance). This proportionality insight is not a matter of arithmetic or geometry. That the relation $v = \alpha \, l/t$ holds is an insight belonging to kinematics, and kinematics belongs to physics; that is, the relation belongs to physics, be it trivial or not.[17]

We can now see that the metre definition apart from a time unit definition also presupposes (a) that the kinematic relation stated by $v = \alpha \, l/t$ is true, and (b) that the unit for velocity is chosen in such a way that $\alpha = 1$. That is, *the purportedly derived unit metres-per-second is already presupposed when the definition of the metre is presented.* From a logical point of view, what magnitudes are to be regarded as coordinated with the concepts "1 metre" and "1 metres-per-second" become simultaneously stipulated in a single coordinative definition.

That from a logical point of view the metre and the metres-per-second magnitudes are simultaneously defined, and so made mutually dependent, is also shown by the fact that one could equally, without any change of substantial content, make the velocity unit instead of the length unit *look like* a base unit. It is the velocity unit that directly gives expression to the unit grounding constancy chosen, the velocity c. As the metre once was defined as a fraction of the meridian, one can define 1 unit velocity to be the fraction $1/299\ 792\ 458$ of the velocity of light; and then make the metre a derived unit by defining it by means of the formula $l = v \, t$ (just as before 1983 the metres-per-second was derived from $v = l/t$).

Before 1983 all base unit definitions for a kind-of-quantity were grounded in a magnitude constancy that belonged to the same kind-of-quantity as the unit magnitude belonged to, but in the metre definition of 1983 this is not so. Here, *the standard unit definition for one kind-of-quantity (length) is grounded in a constancy of another kind-of-quantity (velocity).* Always when this is the case, the first standard unit must in its definition have an explicit or implicit reference to a standard unit of the second kind-of-quantity. Therefore, it is since 1983 logically wrong to call the metre a base unit and the metres-per-second a derived unit.[18]

The New SI retains all the substantial content of the 1983 definition but gives it a new linguistic dress. And in this it becomes quite clear that the metres-per-second unit magnitude is part of the metre definition. What length magnitudes are to be called 1 metre are defined by using (a) the constancy of the velocity of light, (b) the kinematic relation $l = v\,t$ and (c) the velocity unit metres-per-second $(m\ s^{-1})$. The definition proposal looks like this:

> The metre, symbol m, is the SI unit of length. It is defined by taking the fixed numerical value of the speed of light in vacuum c to be 299 792 458 *expressed in the [SI] unit [for speed]* m s^{-1}, where the second is defined in terms of the caesium frequency $\Delta\nu_{Cs}$.
>
> (SI9 Draft, 2016, p. 6, emphasis added)

Despite this open admittance that the metre definition contains the "derived" standard unit for velocity, even the 2013 version of the New SI asserts that all their "description[s] in terms of base and derived units remains valid, although the seven defining constants provide a more fundamental definition of the SI" (SI9 Draft, 2013, p. 2). In particular, as can be seen, it claims that its description of the metre as a base unit and the metres-per-second as a derived unit "remains valid". To me, this conceptualization heavily blurs the normal connotations of the words "base unit" and "derived unit", and I will return to this issue in Section 7. In the long run, this cannot be a good thing. Using $\leftrightarrow$ as a symbol for mutual dependence, and $\downarrow$ for unilateral dependence, what I have claimed can be symbolized thus:

the metre $\leftrightarrow$ the metres-per-second

$\downarrow$

the second

Note that this complaint of mine is only semantic, not epistemic. Both when the metre magnitude is defined by a prototype or as a multiple of wavelengths, and the metres-per-second really is a derived unit, and when as now the metre and the metres-per-second are immediately made mutually dependent, it is the same relation $v = \alpha\, l/t$ that connects the units. That is, both the pre-1983 and the post-1983 SI systems take (rightly of course) this kinematic relation as simply given. To anticipate, in the next two cases I do not find the corresponding physical relation equally unproblematic – in the second case, rather highly problematic.

6.2 The case of the ampere and the coulomb

The present definition of the unit magnitude for the kind-of-quantity electric current is this:

> The ampere [= 1 ampere] is that constant current which, if maintained in two straight parallel conductors of infinite length, of negligible circular

cross-section, and placed 1 metre apart in vacuum, would produce between these conductors a force equal to 2×10^{-7} newton per metre of length.

(SI8, 2006, p. 113)

Explicitly, the definition of the ampere is made dependent on one base unit (the metre) and one derived unit (the newton, which is unit for the kind-of-quantity force). In the same way as the metres-per-second was pre-1983 derived by means of the physical relation $v = \alpha\, l/t$, the newton unit is derived by means of Newton's second law, $F = \alpha\, ma$ ($\alpha = 1$); and since this law only functions in tandem with the first and the third, all three are used. This derivation presupposes units for the base kind-of-quantity mass and the derived kind-of-quantity acceleration; and acceleration presupposes in turn a unit for velocity, which presupposes units for length and time duration. In this stepwise way the present SI looks upon the newton as a unit derived from the base units the metre, the kilogram and the second. As in the former subsection, I would like to stress that such a derivation is not in any sense a purely mathematical derivation. Apart from the base units chosen, Newton's three laws of motion and classical kinematics are presupposed.

The ampere unit magnitude cannot in a similar way be derived from kinematics, mechanics and the connected standard units chosen. A law from classical electrodynamics is needed, and the one used is Ampère's force law: $F_m = 2k_A\, i_1 i_2 / r$. Here i_1 and i_2 are currents in two straight parallel conductors, r is the distance between them, F_m is a variable for the kind-of-quantity force-per-length-unit and $2k_A$ represents (I would say) both a constant of nature and a unit adjuster of the kind I have symbolized α. Surely, the definition of the base unit ampere has many presuppositions, but it does not presuppose a unit for any other electrodynamic kind-of-quantity than that of electric current; in particular it does not refer to a unit for electric charge. The ampere is made the base unit of electrodynamics.

The kind-of-quantity electric charge (q) is in the present SI regarded as a unit derived by means of the electrodynamic relation: $q = \alpha\, i\, t$. In the construal α is set equal to 1, the unit chosen for i is the ampere and the one for t is the second; the unit for electric charge, the coulomb, is then defined as 1 coulomb (C) = 1 ampere (A) × 1 second (s). The definition is often shortened to 1 C = 1 A × 1 s or to C = A s.

In the light of this, let us look at the New SI. Its definition states:

> The ampere, symbol A, is the SI unit of electric current. It is defined by taking the fixed numerical value of the elementary charge e to be 1.602 176 620 8 × 10^{-19} when expressed in the [SI] unit [for electric charge] C, which is equal to A s, where the second is defined in terms of the caesium frequency $\Delta\nu_{\text{Cs}}$.
>
> (SI9 Draft, 2016, p. 7)

The reason behind the proposal is that modern physics takes the elementary charge e (positive in protons and negative in electrons) to be unchangeable and the same everywhere, i.e., e is regarded as a true spatiotemporal constancy. So far, I have no objections, but look at the definition proposal once more. In the very definition of the purported *base* unit ampere, one finds the purportedly *derived* unit coulomb, C. The structure that in the former section was shown to exist between the metre and the metres-per-second via the kinematic relation $v = l/t$, does here hold between the ampere and the coulomb via the electrodynamic relation $q = i\, t$. The electric current magnitudes that are to be called 1 ampere are set by using (a) the constancy of the elementary charge, (b) the electrodynamic relation $q = i\, t$ and (c) the electric charge unit coulomb.

From a logical point of view, the ampere unit and the coulomb unit are, in the New SI, simultaneously chosen and defined in a way that is analogous to that between the metre unit and the metres-per-second unit (in both the present and the New SI). Therefore, I find it here equally semantically misleading to talk about a distinction between a base unit and a derived unit. There is, though, an epistemic difference; the relation $q = \alpha\, i\, t$ might be easier to contest than $v = \alpha\, l/t$.

That the ampere and the coulomb are simultaneously defined is also shown by the fact that one could just as well have started by defining the coulomb unit. It is the coulomb unit that directly gives expression to the unit grounding constancy chosen, the elementary charge e. There are no problems in defining 1 unit electric charge (the coulomb) as being equal to the multiple $1.602\ 176\ 620\ 8 \times 10^{-19}$ of the elementary charge, and then make the ampere a derived unit by defining it by means of the relation $i = e/t$ and the formula $1\ \text{A} = 1\ \text{C/s}$ (just as now the coulomb is derived from the ampere by means of $1\ \text{C} = 1\ \text{A} \times 1\ \text{s}$).

Using $\leftrightarrow$ as a symbol for mutual dependence, and $\downarrow$ for unilateral dependence, what I have claimed in this subsection can be symbolized thus:

the ampere $\leftrightarrow$ the coulomb

$\downarrow$

units belonging to kinematics and mechanics

Both the present SI and the New SI take a number of physical relationships as simply given when they define the ampere and the coulomb, and most of these relationships are not as indisputable as the relation between velocity, length and time is. However, it is exactly the same relationships that are presupposed in the present SI and the New SI. Therefore, there is no epistemic difference between how the two definitions handle the ampere and the coulomb.

Before turning to the next case, I would like to highlight a special feature of the present definition of the ampere; I need to refer to it in the next subsection.

When unit definitions are anchored in theories instead of in obviously existing macroscopic magnitudes such as prototypes, meridians and solar days, then quite a special metrological possibility arises. Since theories are not only about what actually exists and can exist, but also about counterfactual situations that may never exist and may not even possibly exist, the constancy magnitude that grounds a unit definition can be placed in a non-existing or impossibly existing magnitude.

This is the case in the present definition of the ampere. The definition contains the phrase "two straight parallel conductors of infinite length", but such conductors are physically impossible. As noted by Tal (2011, p. 1087), the same is true of the definition of the second. It refers to a kind of atom that has the impossible temperature state of zero degree kelvin.

6.3   *The case of the kilogram and the joules-times-seconds*

First some words about what triggered the opinion that the kilogram prototype has done its duty and should go. This is in order to substantiate my intimation in Section 1, that I find it a bit remarkable how contemporary philosophy of science has neglected the problem of constancy. I start with a quotation:

> A pivotal event took place in 1988, when the IPK [the international prototype kilogram] was removed from its safe and compared with the six identical copies kept with it, known as *témoins* (witnesses). . . . The verification in 1988 confirmed this trend: not only the masses of the témoins but those of practically all the national copies had drifted upward with respect to that of the prototype. . . . Quinn, who became the BIPM's director in 1988, outlined the worrying implications of the apparent instability of the IPK in an article published in 1991. Because the prototype *is* the definition of the kilogram, technically the témoins are gaining mass. But the "perhaps more probable" interpretation, Quinn wrote, "is that the mass of the international prototype is falling with respect to that of its copies"; that is, the prototype itself is unstable and losing mass.
>
> (Crease, 2011, pp. 253–255)

Whether this decision should be reckoned "*Fingerspitzengefühl*" or merely a kind of metrological common sense, I am not the man to tell, but that is unimportant. Here is a quotation from a philosophical book on the foundations of science, published in 1919:

> If we have made many copies of a unit and found, just after they were made, that they and the unit were all equal, and if we find later that the copies and the unit are not still equal, then we can say either that the copies have changed or that the unit has changed. If all the copies (or nearly all) are still equal to each other, though differing from the unit, then we shall say that the unit has changed.
>
> (Campbell, 1957, p. 362)

This kind of sensitivity to the problem of constancy is mostly absent or abstracted away in the philosophy of science after 1950. There is then only a simple noticing that a standard unit is needed and has to be chosen. And Carl G. Hempel is in his classic treatise on scientific concept formation using the very kilogram prototype as his example. Without any qualifications at all, he simply states: "A specific object k, the International Prototype Kilogram, is to serve as a standard and is to be assigned the m-value 1,000" (Hempel, 1952, p. 63).

The kilogram definition of the New SI looks like this:

> The kilogram, symbol kg, is the SI unit of mass. It is defined by taking the fixed numerical value of the Planck constant h to be $6.626\,070\,040 \times 10^{-34}$ when expressed in the [SI] unit [for action] J s, which is equal to kg m² s⁻¹, where the metre and the second are defined in terms of c and $\Delta\nu_{Cs}$.
>
> (SI9 Draft, 2016, p. 6)

This means that what mass magnitudes should be called 1 kilogram are set by using (a) the Planck constant, (b) some physical relationships and (c) the standard unit for action, joules-times-seconds (J s). When the kind-of-quantity action is given the unit J s, it is looked upon as being energy *multiplied* by time; compare velocity, which is length *divided* by time. In many contexts it is expedient to call both multiplication and division multiplications (using the general rule $1/x^n = x^{-n}$; e.g., m/s = m s⁻¹), but in metrological contexts such a wide multiplication notion can hide important things (see remark V).

I have pointed out that the metre definition also simultaneously defines the velocity unit metres-per-second, and that the New SI's ampere (A) definition simultaneously defines the charge unit coulomb (C). In the same way, the New SI's kilogram (kg) definition simultaneously defines the action unit joules-times-seconds (J s). Compare the end of the ampere definition, C = A s, with the end of the kilogram definition, J s = kg m² s⁻¹. In the metre case, the physical relationship that makes a two-sided coordinative definition possible is a basic purely kinematic relationship, and in the ampere case the corresponding relationship is purely electrodynamic, but what does the relationship look like in the kilogram case?

For some years I thought for good reasons (see remarks I and III) that the metrologists behind the New SI had found the relationship wished for in $m = h\nu/c^2$, a formula that is derivable in quantum electrodynamics. But things have changed. In the 2013 draft and in the 2016 draft this formula is no longer mentioned. I am well aware of the fact that a good thing can be proposed without arguments, and even be defended by bad arguments, but the way the proposed kilogram definition has been argued for between 2010 and 2016 has made me suspicious. In fact, I have for some time and my own fun used the expression "the mistake of falling in love with an analogy inference" to characterize what I suspect is operative beneath the different argumentations put forward.

The analogy inference I have in mind runs as follows: (a) c and h are the two truly fundamental constants in contemporary physics, (b) the metre and the kilogram are beside the second the most fundamental units in the SI system, (c) the metre is defined by means of c, therefore, analogously, (d) let us defines the kilogram by means of h. Briefly put, what c is to the metre, h must be to the kilogram. Analogy inferences have played quite an important role in the development of science, but if there are good arguments against their validity, they should be forgotten.

I will make six remarks connected to the arguments that have been used to support the kilogram definition of the New SI. Taken together, the remarks have convinced me that the definition is much less reasonable than an alternative definition that has been proposed by some other metrologists, and which I will present in the sixth remark. This latter definition is not contaminated by any of the obscurities that I find in the arguments for the New SI definition.

Remark I

In the first New SI draft, it is said that the two famous equations $E = mc^2$ and $E = hv$ "together lead to $m = hv/c^2$" (SI9 Draft, 2010, p. 7). Hereby, the kind-of-quantity mass as used in relativity theory and the kind-of-quantity action from quantum mechanics become directly connected. In this operation it is taken for granted that the energy variables of relativity theory and of quantum mechanics are variables for the same kind-of-quantity. Otherwise the equations $E = mc^2$ and $E = hv$ cannot immediately be combined into the desired relationship.

As far as I know, there is still no theory, superstring theory or overarching relativistic quantum-mechanical theory, which has managed to combine the relativistic mass of relativity theory with the rest mass of much of quantum mechanics and quantum chemistry. It seems to me as if the New SI metrologists, for a while, were trying to synthesize where physicists have not yet been able to do so; the view presented is not to be found in the later drafts.

Remark II

However, even if the deleted derivation had been valid, the logical structure of the kilogram definition would *differ in complexity* from that of the metre and the ampere definitions. The central equations used in the latter definitions ($v = l/t$ and $q = i\, t$) contain *three* variables each, and only a specification of t is needed in order to obtain the needed direct mutual connection between the other two kind-of-quantity variables. In the formula $m = hv/c^2$, on the other hand, there are *four* kinds-of-quantity involved. That h and c are constants, not variables, does not change this fact.

As c and h appear in the formula, they are no more than the variables m and v tied to any specific measurement units (see subsection 6.1). They are, however, tied to specific kinds-of-quantity just as much as m and v are. A fundamental constant represents *an unchanging magnitude of a certain kind-of-quantity*. The constant c can be stated in metres-per-second, kilometres-per-hour, yards-per-minute and many others, but all these units have to be units for velocity. Similarly, the Planck constant is a constant magnitude of the kind-of-quantity action, whatever standard unit it is ascribed; in the SI it is joules-times-seconds. As the symbol c represents a constant velocity magnitude, the symbol h represents a constant action (or difference of action) magnitude.[19]

This being so, the question one could have asked is: how does the New SI choose the specific frequency v needed for the definition to work? What we find stated is a relationship that can be rewritten as: 1 (kg) = 1.475 521 . . . × 10^{40} hv_{Cs}/c^2 (SI9 Draft, 2013, p. 13). Here, v_{Cs} represents a frequency (in hertz) of a radiation that occurs in certain energy level transitions in caesium 133 atoms, and which always has the same frequency;[20] h and c are the usual constants expressed in SI units. What is to be regarded as 1 kg rest mass is defined by a mass multiple (1.475 521 . . . × 10^{40}) of the three constancies of nature: h, c and v_{Cs} as related in hv_{Cs}/c^2.

Of importance in relation to the following remark III is that the frequency used (v_{Cs}) is not a de Broglie-Compton frequency, but a radiation frequency.

Remark III

To repeat: the first New SI draft was presented in September 2010, and a later draft was presented in December 2013, but an intermediate specification was presented in March 2013. In order to let the m in $m = hv/c^2$ represent also rest mass, the authors behind this specification rely on de Broglie's hypothesis that not only particles that lack rest mass (photons) conform to the equation $E = hv$, but that even particles with rest mass do. There are not only radiation waves, there are also, to use the old expression, *matter waves*, waves whose frequency is proportional to the rest mass of the particle in question:

> Note that according to the theories of special relativity and quantum mechanics, an atomic particle of mass m, and hence from the Einstein relation of total energy $E = mc^2$, may be interpreted as a wave with an oscillation frequency, called the de Broglie-Compton frequency, given by $v = mc^2/h$.
> (DraftCh2_4March2013, p. 8; never made public)

On this view, for particles at rest the variable m represents the rest mass, not a relativistic mass. In other words, the variable v used does not represent radiation frequencies but de Broglie-Compton frequencies.

Now a rhetorical question: can the existence of matter waves be regarded as so ascertained that they should be allowed to become an essential part of a metrological kilogram definition? Note that even in the quotation there is some hesitancy. It is said that the mass of an atomic particle "*may be interpreted* as a wave with an oscillation frequency" (emphasis added). Shouldn't it, from a metrological point of view, be beyond all present reasonable doubt that all atoms, molecules and macromolecules have an oscillation frequency?

Moreover, if the authors had really trusted their view that the kilogram definition relies on de Broglie-Compton frequencies, they could have reasoned much more directly. Let me for a moment play with $m = hv/c^2$ on the assumption that in fact all kinds of atoms, molecules and macromolecules have a matter wave with a de Broglie-Compton frequency.

On this assumption it is possible to use de Broglie-Compton frequencies the way electric current is used in the present ampere definition. One may then speak of specific non-existing matter waves the way the ampere definition speaks of electric currents in non-existing conductors of infinite length. That is, one may then speak of a de Broglie-Compton frequency that is equal to the radiation frequency ν_{Cs}. One can simply substitute ν_{Cs} by the corresponding de Broglie-Compton frequency and so obtain a definition of a *rest mass magnitude* of 1 kg.

Even more, one can put in a frequency that directly corresponds to a rest mass of 1 kg. There is no longer any reason to put in a frequency that corresponds to a rest mass that is only a fraction of 1 kg. The present metre definition can be formulated thus: 1 metre is the distance travelled in 1 second by light in a vacuum when its velocity is given the value 299 792 458 m/s. A similar formulation of the content of the proposed kilogram definition looks like this: 1 kilogram is the rest mass of a particle whose matter wave frequency in hertz is such that the Planck constant is equal to $6.626\ 070\ 040 \times 10^{-34}$ J s.

If in the formula under discussion we put in the values for c and h in SI units, and put in 1 kg on the left hand side, then we can easily calculate a de Broglie-Compton definition frequency (ν_d) for 1 kg. That is, by means of the equation $1 = h\nu_d/c^2$ we can find the de Broglie-Compton frequency that shows that something has the rest mass 1 kg. The rounded number of ν_d is 1.4×10^{50} hertz. As far as I know, no physicist thinks that *aggregates* of molecules can be ascribed a certain de Broglie-Compton frequency. Otherwise the frequency mentioned could lay claim to be the the de Broglie-Compton frequency of the existing kilogram prototype.

The purpose of my last comments is to show, that if the authors referred to had found the existence of matter waves to be truly unproblematic, then in all probability they would themselves have formulated the views I have just played with. And I take the fact that de Broglie-Compton frequencies are not mentioned in later drafts as a sign that, today, at most a minority of the people behind the issuing of the New SI think that the notion of matter waves is needed to back the kilogram definition.

Remark IV

In the drafts of 2013 and 2016, the relationship $m = h\nu/c^2$ is not mentioned at all. Instead the watt balance (since June 2016 also called Kibble balance) is given a prominent position, and I will briefly give some comments on the way it helps to create a numerical connection between m and h. First the 2013 draft:

> Note that macroscopic masses can be measured in terms of h, using the Josephson and quantum-Hall effects *together with* the watt balance apparatus.
> (SI9 Draft, 2013, p. 14, emphasis added)

The central equations of the Josephson and the quantum Hall effects contain the Planck constant but no variable for mass. The variables in question are all electrodynamic variables, which then by means of other relationships can be connected to a variable for mass. This means that in the 2013 proposal for the kilogram definition it is impossible to find any direct relationship that fulfils the role that $v = l/t$ has in the metre definition and $q = i\,t$ has in the ampere definition. In this respect, the kilogram definition is now backed by an argumentation whose logical structure differs radically from that of the metre and the ampere definitions. And this structure is retained in the 2016 version, even though now a numerical relationship that contains both *m* and *h* is in fact explicitly presented:

> For the kilogram, the unit whose definition has undergone the most funda-mental change, realization can be through any equation of physics that links mass, the Planck constant, the velocity of light and the caesium frequency. One such equation is that which describes the operation of an electro-mechanical balance, known as a watt balance. With this a mechanical power, measured in terms of a mass, *m*, the acceleration due to gravity, *g*, and a veloc-ity, *v*, can be measured in terms of an electrical power measured in terms of an electric current and voltage measured in terms of the quantum Hall and Josephson effects respectively. The *resulting equation* [emphasis added] is $mgv = Ch$ where *C* is a calibration constant that includes measured frequen-cies and *h* is Planck's constant.
>
> (2016, draft, p. 27)

The equation that connects *m* and *h* is rightly called a *resulting equation*. It is not a basic equation in any physical theory. It is an equation that captures the essential functions that a watt balance can be put to. This ingenious device can, for instance, (as hitherto) be used to experimentally determine the numerical value of *h* when mass has a standard unit that is fixed independently of *h*; or, conversely, be used (if the New SI is accepted) to determine the values of certain masses by means of *h*. If the New SI is accepted, the particular watt balance that is regarded as being the most accurate one will in all probability function as what might be called a *prototype realizer* for the kilogram; the other watt balances (and other similar devices) will then function as so-called *témoins* (see the first quotation in subsection 6.3).

By stating this, I am neither saying nor believing that there is something wrong with the watt balance as such. The question I am posing, and myself answering negatively, is whether at the moment *metrology* really benefits from making *h*, and thereby the watt balance, central to the kilogram definition.

The symbol *h* is meant to represent a fundamental constant of nature, not just an atomic parameter that is invariant throughout time and space. Therefore, I think that more discussion is needed about of what kind of constancy it really represents; what *h* represents cannot be taken to be as known as what *c* represents is. The next remark is meant to make this view of mine more concrete.

Remark V

In geometry, no one would neglect the distinction between points and lines, but when metrologists are discussing action they neglect the distinction between time points and time durations (lines). Let me explain my charge.

Spatial lines can have properties that points cannot, and vice versa; lines and points are in this sense radically different. For instance, lines can be curved or straight, and two straight lines are either parallel or not, but points cannot possibly have any of these properties. Let us now think of energy. A certain magnitude of energy can (the uncertainty relation of quantum mechanics disregarded) be ascribed to a particle both at specific single points of time and, if constant, to a time duration. So far, no point-versus-line-problem.

Let us next think of energy changes (ΔE). They cannot possibly take place at a single point of time, even though a measure of energy change *velocity* (dE/dt) can be defined for time points. Energy change velocity has in the SI system the measurement unit joules-*per*-second (J s^{-1}). Action, on the other hand, has the unit joules-*times*-seconds (J s), and here the old mathematical distinction between division and multiplication becomes important. Values of J s^{-1} can be ascribed to something at a time point, but values of J s cannot; the energy (or energy difference) measured by J can in the latter case be ascribed only to something with a temporal duration.

The fact that action should be regarded as energy times time duration is clearly visible in the classical mathematical definitions of action (S), where action is defined as an integral between two different time points, t_1 and t_2, as in $S = \int L \, dt$. If the points are identical, then, necessarily, an integration of L gives the result zero; that is, there can be no action at a single point of time.

In the sense explained, it is as impossible to think of an action as existing at a single point of time, as it is impossible to think of a volume as existing in a plane, a surface existing in a line or a line in a point. This applies both to the pre-quantum-mechanical notion and to the quantum-mechanical notion of action. In both, action is an attribute of something that requires an interval of time for its existence; be it a continuous or a quantized attribute.

The fact now made visible is eradicated by the way many metrologists and physicists allow themselves to use the equality J s = kg m^2 s^{-1}, which, note, is an explicit part of the kilogram definition (see earlier quotation). The equality is wrongly used as if the left and the right hand sides are always substitutable for each other, but they are not. I claim that J s can be a unit only for something that is time-extended, but I do by no means make a similar claim for the unit kg m^2 s^{-1}. It can be applied to magnitudes of kinds-of-quantity at a single point of time; and is rightly so applied when talking about pre-quantum-theoretical angular momentum. In my opinion, whatever good use the expression J s = kg m^2 s^{-1} can be put to, it cannot possibly be used to show that the symbols on the left and the right hand sides always can refer to the same kind-of-quantity. This view of mine comes in conflict with the following view of the present SI system:

> Symbols for units are treated as mathematical entities. In expressing the value of a quantity as the product of a numerical value and a unit, both the numerical value and *the unit may be treated by the ordinary rules of algebra.*
>
> (SI8, 2006, p. 132; emphasis added)

The SI view allows one wrongly to reason as follows: action has the unit J s; however, since J equals kg m² s⁻², J s is equal to (kg m² s⁻²) s, which, in turn, is equal to the unit kg m² s⁻¹, because (s⁻² s) = s⁻¹. Now, if s is a variable for pure real numbers, the last equality is of course true, but s does here represent a metrological unit, the second, which is quite another thing. In conformity with some other critics, especially the metrologist Walter Emerson (2004a, 2004b, 2008), I have earlier demonstrated why metrological multiplications and divisions do not allow for all the operations that arithmetic multiplications and divisions allow (Johansson, 2010, part II).

Of particular interest now is that – contrary to the SI view – it makes no physical sense to *divide* a base unit with itself. For instance, saying that s/s = 1 requires the introduction of a special *measurement unit* called "one"; since the numeral 1 on the right hand side cannot represent a pure number. In the critical writings referred to, it is argued that such a unit one is not needed; and, even worse, it creates confusion. Emerson's and my rejection of the unit equality s/s = 1 also means that I find the unit equality (s⁻² s) = s⁻¹ unacceptable, which, in turn, keeps intact my view that J s, unlike kg m² s⁻¹, can be a measurement unit only for something that has a temporal duration.

Most physical and chemical kinds-of-quantity can in a commonsensical way be regarded as properties of particles, waves, statistical ensembles, movements or processes.[21] Even though action can be called a property in a wide sense of this term, it cannot, as I have shown, be a property that can be ascribed to something at a single point of time, which the kinds-of-property length, mass, velocity, energy, electric charge and (intensity of) electric current can. Conclusion: the constancy of nature that is reflected in the notion of the Planck constant cannot be a property constancy of the kind exemplified by the constants *c* and *e*.

The special feature of action that I have highlighted – necessarily being time-extended – is never mentioned and remarked upon in any of the drafts of the New SI, but I think it is such a peculiar feature that until the nature of action has been better clarified, metrologists should be extremely reluctant to make the Planck constant central to a base unit definition.

Remark VI

The concrete kilogram prototype in Sèvres can be substituted in other ways than by the definition proposed in the New SI drafts. As a number of metrologists have pointed out, the kilogram magnitude can also be defined by means of the constant mass of a kind of atom. A specific such proposal was already a year after the appearance of the first draft of the New SI put forward as being a better alternative, namely this one:

A kilogram is the mass of 84 446 889³ × 1000/12 unbound atoms of carbon-12 at rest and in their ground state.

(Hill, Miller, & Censullo, 2011, p. 84a)

Here, the kilogram magnitude is a multiple of the mass of a certain kind of atom, i.e., a specified atomic parameter. In logical structure, the definition is completely analogous to the SI definition of the metre between 1960 and 1983, when the metre was defined as a multiple of a certain wavelength of a certain radiation.

In Section 2, I reported that in the last draft of the New SI the earlier general talk of "constants of nature" has been replaced by the simple generic term "defining constants", and that these constants are said to be of four different kinds: fundamental constants of nature, specified atomic parameters, conversion factors and technical constants. This means that one of the original reasons in favour of the Planck constant based definition of the kilogram has now disappeared. The reason was that all the seven base units should be defined by means of fundamental constants of nature, but this view is no longer adhered to.

I will next briefly bring in the definition of the second. It is the truly basic base unit in the New SI. It is presupposed in the new definitions of all the other base units except the mole. Here is the definition:

> The second, symbol s, is the SI unit of time. It is defined by taking the fixed numerical value of the caesium frequency $\Delta\nu_{\text{Cs}}$, the unperturbed ground-state hyperfine transition frequency of the caesium 133 atom, to be 9 192 631 770 when expressed in the unit Hz, which is equal to s^{-1}.
>
> (SI9 Draft, 2016, p. 5)

This definition does not bring in any fundamental constants of nature. Instead it uses a specification of an atomic parameter, just as the alternative definition of Hill, Miller and Censullo does. Therefore, the use of an atomic parameter in the alternative definition can no longer be regarded a reason against it. And the alternative definition is much simpler than that of the New SI.

Conclusion: it must be more reasonable to make the ninth edition of the SI system contain the kilogram definition proposal of Hill, Miller and Censullo – or something similar – than to insert the one proposed in the New SI drafts.

I can't get rid of the suspicion I have ventured: many of those who support the New SI definition of the kilogram seem to be seduced by a thrilling and beautiful analogy inference.

7   General conclusion of Sections 5 and 6

As already quoted in Section 5, the present SI says: "It should be recognized that . . . respective base units – the metre, kilogram, second, ampere, kelvin, mole, and candela – are in a number of instances interdependent" (SI8, 2006, p. 111). In Section 6, I have shown that in several cases there are interdependencies between base units and derived units, too.

Now, if the definitions of the base units are not always independent of each other, and even in some cases dependent on the definitions of derived standard units, then I would say that common sense semantics requires that the SI division of the standard units into groups called "base units" and "derived units", respectively, should be deleted. Instead, all the implicit dependence relations among the unit definitions, sometimes hinted at, could be clearly stated. I can see no conclusive reason not to make the definitional structure of the SI system transparent. As I have made clear in relation to the kilogram definition, there have been changes in how this definition has been argued for in the different drafts. Similarly, what the distinction between base and derived units amounts to, has been given different presentations in the different drafts. The last draft, speaking as if it has become accepted, says as follows:

> Prior to the 2018 definitions, the SI was defined through seven *base units* from which the *derived units* were constructed as products of powers of the *base units*. Defining the SI by fixing the numerical values of seven defining constants has the effect that this distinction is, in principle, not needed, since all units, *base* as well as *derived units*, may be constructed directly from the defining constants. Nevertheless, the concept of base and derived units is maintained, not only because it is useful and historically well established, but also because it is necessary to maintain consistency with the International System of Quantities (ISQ) defined by ISO/IEC 80 000 series of Standards which specify base and derived quantities to which the SI base and derived units necessarily correspond.
>
> (SI9 Draft, 2016, p. 5)

I think this retention of the distinction is a bad decision. Retaining what is "historically well established" but in fact "not needed", hides the radical shift in metrological thinking that has taken place between the present and the New SI. It is also for some unknown reason claimed that the distinction "is useful", but I think, to the contrary, that it will create much confusion among many metrologists, scientists and philosophers of science. The required "consistency" with other metrological documents could easily have been solved by some appendices.

8 The possibility of metrological improvements

Is it possible to test whether or not an assumed unit-grounding constancy C_1 really is constant? Doesn't that require that we compare it with another assumed constancy C_2? And then we can of course ask how C_2 is to be tested, and so on into C_n and an infinite regress. Now, if C_1 and C_2 are nothing but spatiotemporally individual objects, then such an infinite regress does arise. However, as shown by the story of the kilogram prototype and its copies, not even a prototype is in fact treated as nothing but an individual object. It is in effect regarded as one of a number of exactly similar objects; and it can (at least in principle) be tested whether these particular objects change in relation to each other. If they do, something must be wrong somewhere. Thus, testing constancy need not involve an infinite regress.

When prototypes are exchanged for properties of atoms or electromagnetic radiation, then there is no longer any need to distinguish between prototype and copy. Each light ray in a vacuum is so to speak both prototype and copy, and the same goes for the elementary charge. Therefore, it is in principle possible to try to test whether their assumed constancy holds or not. However, in the New SI nothing is said about how to check the constancies it uses to ground the web of dependencies that exist between the units chosen. In this sense, it does take on the look of a Zanzibar system. The assumed constancy of the kilogram prototype was meant to be checked once in about every 40 years, even though after 1889 there were only two checks (1946 and 1989–91), but when and how are the presumed constancies of the New SI to be tested? No one knows.

Some physicists do direct measurements in order to detect whether assumed constants really can be considered constant. Perhaps the most well-known are those concerned with the fine structure constant. But there is also another but indirect way in which lack of constancy can be detected. If the theory in which the constancy appears meets empirical anomalies, i.e., reliable measurements that do not fit what the theory predicts, then automatically the presumed constancy becomes questioned, too. For some months some years ago there seemed to be such a case, which is enough for my pedagogical purposes.

In September 2011, physicists within the so-called Opera collaboration released results, which they claimed showed that neutrinos can move faster than light, but already in February 2012 they withdrew the claim. Now, if instead their initial beliefs would have been repeatedly confirmed, something would have had to be changed in relativity theory. Of course, whether or not the best change would be to say that c is after all not a constant is an open question. Nonetheless this "neutrino case" illustrates the possibility of concluding that c is not a constancy of nature. And the same unspecific possibility exists with respect to h. Not even the inevitable epistemic circularities mentioned in Section 4 make physics and chemistry completely cut off from the possibility of meeting recalcitrant measurement data; data that later in retrospect may take on the appearance of falsifications.

Famously, in his *Philosophical Investigations* Ludwig Wittgenstein makes a metrological remark:

> There is *one* thing of which one can say neither that is one metre long, nor that it is not one metre long, and that is the standard meter in Paris. – But this is, of course, not to ascribe any extraordinary property to it, but only to mark its peculiar role in the language-game of measuring with a metre-rule.
> (Wittgenstein, 1967, §50)

So far so good, one might say. Surely, nature had not given the standard metre the property one metre or any other numerical property. It was a community of human beings that by means of a coordinative definition ascribed the numerical concept "1 metre" to a property instance of a platinum-iridium cylinder in Sèvres outside of Paris. But what Wittgenstein forgets is the fact that – very importantly – the standard metre was assumed not to change its length. This is, I would say, to ascribe a macroscopic thing an "extraordinary property". Wittgenstein's philosophy contains no stress on the necessity of making changes in a language game, but metrology needs changes now and then. Therefore, I would like to end by paraphrasing a famous passage from the philosopher of science Otto Neurath (the words within square quotes are his):

> [We] Metrologists are like sailors who on the open sea must reconstruct their ship but are never able to start afresh from the bottom. Where a [beam] unit definition is taken away a new one must at once be put there, and for this the rest of the [ship] metrological system is used as support. In this way, by using the old [beams and driftwood] units and new natural-scientific knowledge the [ship] metrological system can be shaped entirely anew, but only by gradual reconstruction.
>
> (after Neurath, 1973, p. 199)[22]

Acknowledgements

This chapter owes much to the straightforward discussions that have taken place in a couple of small mailing groups concerned with metrology, where, being a philosopher, I was lucky to become accepted. Also, I am quite indebted to Olivier Darrigol for forcing me to make subsection 6.3 as clear as (hopefully) it now is.

Notes

1 For a check, look for instance at the Carl G. Hempel classic (1952, ch. 12), via the otherwise extraordinarily rich book by Henry E. Kyburg (1984), to recent set-theoretical papers such as Frigerio, Giordani, and Mari (2010), Giordani and Mari (2012) and Rossi and Crenna (2013). A small exception is some discussions around Hans Reichenbach's notion of *coordinative definition* (Reichenbach, 1958, §4).

2 I am not, however, the only contemporary philosopher of science who tries to change the situation; see e.g. Hasok Chang (2004) and Eran Tal (2011).

3 The term "standard unit" cannot be found in either the *SI Brochure* (SI8, 2006) or the terminological clarifications in (VIM3, 2012), but it is the natural term to use as a common label for base units and derived units. Both kinds of units are measurement units (VIM3, 2012, def. 1.9), but so are also multiples and submultiples of them.

4 In my article (Johansson, 2014), I discuss the mole and the second, too.

5 The *SI Brochure* does not contain the term "kind of quantity", but for terminological clarifications it refers to VIM3 (SI8, 2006, p. 103); and VIM3 contains a definition of the term (VIM3, 2012, def. 1.2). The hyphenation is not in VIM3, but it is not an invention of mine. I have taken it from the distinguished metrologist René Dybkaer (2009), who intends his book to be in conformance with VIM3.

6 As I have pointed out elsewhere (Johansson, 2010, pp. 220–221), the SI can be given both a realist and a nominalist reading of its property talk; therefore, I see no reason to bring in the realism–nominalism issue here.

7 In the philosophy of science, the necessity of such realizations is stressed by Tal (2011).

8 My notion of "epistemic circularity" must by no means be conflated with the notion of "epistemic iteration" put forward by Chang (2004). In an epistemic

iteration, one epistemic circularity is exchanged for another and presumably better one. Chang's views conform well to the claims I make in the concluding section, "The possibility of metrological improvements".

9   In 1799 it was exchanged for a prototype, which, in turn, 1889 was replaced by the present international prototype of the kilogram.

10 The existence of this possibility is in conformance with the following statement: "[A] unit is simply a particular example of the quantity concerned" (SI8, 2006, p. 103). Fractions and multiples can be examples, too.

11 I disregard the terminology according to which "velocity" is used only for vectors and "speed" only for scalars; I treat them as synonyms.

12 If the definition of the second is inscribed, then the metre definition becomes the following: the metre is the length of the path travelled by light in vacuum during a time interval of 30.6633189884984 caesium-133 hyperfine transition cycles.

13 The hyphenation is added in order to make all unit concepts one-word concepts.

14 It must not be read as a case of multiplication! It *cannot* be turned into $\{Q\} = Q/[Q]$; see my article (Johansson, 2010, sect. 5). As far as I can judge, this symbolism is running out of fashion, but it is retained in passing in the New SI (SI9 Draft, 2016, p. 3).

15 This means that we are assuming so-called *ratio scales*. Another kind of metric scale is the linear interval scale that was used by both Celsius and Fahrenheit. This scale, however, is given no place in the SI system. For a presentation of different kinds of scales, see Dybkaer (2009, ch. 17).

16 What has been stated is true only when one looks at an isolated natural law. If the same variables occur in more than one law, then if α is set equal to 1 in one law, it might be impossible to make it equal to 1 in all the others. For instance, as pointed out by Peter Simons (2013, pp. 520–521), if we turn Newton's second law $F = \alpha\, ma$ into $F = ma$, then it is impossible to turn his gravitational law $F = \alpha\, m_1 m_2/r^2$ into $F = m_1 m_2/r^2$. It becomes the well-known $F = G\, m_1 m_2/r^2$ (or $F \propto m_1 m_2/r^2$). If, on the other hand, G is set equal to 1, then $F = \alpha\, ma$ cannot be reduced to $F = ma$. Formally, it is possible to start to regard G as a variable instead of a constant, but *within Newtonian mechanics* it is only a proportionality constant and a unit adjuster. In other words, the "Big G" of Newtonian mechanics is not a constant of nature. This fact is not always made clear.

17 Whether, after relativity theory, the relation should be confined to the inside of an inertial frame of reference, I will not discuss. Once it was self-evident that if one body moves away from you and your inertial frame with the velocity v_1, and a second body moves away from the first in its inertial frame with the velocity v_2, then the second moves away from you with the velocity $v_1 + v_2$. According to the theory of special relativity, however, the second moves away from you with the velocity $v = (v_1 + v_2)/(1 + v_1 v_2/c^2)$.

18 What I have said was at least once noted in the discussions before the 1983 (October) decision, but the observation seems to have had no impact at all. I quote: "The question may well arise as to whether, in view of this, we can continue to regard velocity as being a derived unit. Thus it could be that we might choose to regard velocity as being defined simultaneously with length . . . or even that velocity rather than length was the base unit" (Petley, 1983, p. 375).

19 The parenthesis "(or difference of action)" is needed since action and difference of action have the same measurement unit; this is also true of length and difference of length, etc.

20 What I write as ν_{Cs} is in the draft mentioned symbolized by $\Delta\nu(^{133}Cs)_{hfs}$; hfs is short for hyperfine splitting, and D symbolizes that ν originates from an energy level difference.

21 Of course, this is only the case when they are regarded as referring to something real, not when regarded as only calculation-simplifying mathematical tools. However, if action is only regarded as a mathematical concept, h cannot be regarded as representing an invariant of nature.

22 Very early, Neurath subscribed to the views I have called inevitable semantic and epistemic circularities, which only much later became the mainstream views in the philosophy of science. The quotation stems from Neurath's German book *Anti-Spengler* (1921). In 1932/33 he re-used the metaphor, but in a shortened version that perhaps is more quoted (Neurath, 1959, p. 201). The whole history of his metaphor is described in Thomas Uebel's "On Neurath's Boat" (Cartwright et al., 1996, part 2). Neurath was an ardent spokesman of logical positivism, but nonetheless he never shared two of the views that commonly are regarded as central to this anti-metaphysical movement: (a) that somewhere there are completely theory-free empirical data; and (b) that atomic sentences can be directly compared with reality. The solution to the puzzle is that Neurath, despite his holistic views, believed a scientific and a naturalistic everyday language can be completely cut off from all metaphysics. Here I diverge from Neurath; I think, like Popper, that there is some overlap.

References

Campbell, N. R. (1957). *Foundations of science: The philosophy of theory and experiment*. New York, NY: Dover Publication.

Cartwright, N., Cat, J., Fleck, L., & Uebel, T. (1996). *Otto Neurath: Philosophy between science and politics*. Cambridge: Cambridge University Press.

Chang, H. (2004). *Inventing temperature: Measurement and scientific progress*. Oxford: Oxford University Press.

Crease, R. P. (2011). *World in the balance: The historic quest for an absolute system of measurement*. New York, NY: W.W. Norton.

Dybkaer, R. (2009). *An ontology on property for physical, chemical, and biological systems*. (ISBN 978-87-990010-1-9). Retrieved 19 November 2017 from http://ontology.iupac.org/.

Emerson, W. H. (2004a). One as a 'unit' in expressing the magnitudes of quantities. *Metrologia, 41*, L26–L28.

Emerson, W. H. (2004b). On the algebra of quantities and their units. *Metrologia, 41*, L33–L37.

Emerson, W. H. (2008). On quantity calculus and units of measurement. *Metrologia, 45*, 134–138.

Frigerio, A., Giordani, A., & Mari, L. (2010). Outline of a general model of measurement. *Synthese, 175*, 23–149.

Giordani, A., & Mari, L. (2012). Property evaluation types. *Measurement, 45*, 437–452.

Hempel, C. G. (1952). *Fundamentals of concept formation in empirical science*. Chicago, IL: The University of Chicago Press.

Hill, T. P., Miller, J., & Censullo, A. C. (2011). Towards a better definition of the kilogram. *Metrologia, 48*, 83–86.

Johansson, I. (2010). Metrological thinking needs the notions of *parametric* quantities, units and dimensions. *Metrologia, 47*, 219–230.

Johansson, I. (2014). Constancy and circularity in the SI. *Metrologybytes.net (OP EDs)*. Retrieved 19 November 2017 from www.metrologybytes.net/.

Kyburg Jr., H. E. (1984). *Theory and measurement*. Cambridge: Cambridge University Press.

Metrology Bytes. (2012). Retrieved 19 November 2017 from www.metrologybytes. net/.

Neurath, O. (1959). Protocol sentences. In A. J. Ayer (Ed.), *Logical positivism* (pp. 199–208). New York, NY: The Free Press.

Neurath, O. (1973). *Empiricism and sociology* (M. Neurath & R. S. Cohen, Eds.). Dordrecht: D. Reidel Publishing Company.

Petley, B. W. (1983, June 2). New definition of the metre. *Nature, 303,* 373–376.

Price, G. (2010). Failures of the global measurement system. Part 1: The case of chemistry. Part 2: Institutions, instruments and strategy. *Accreditation and Quality Assurance, 15,* 421–427, 477–484.

Price, G. (2011). A skeptic's review of the New SI. *Accreditation and Quality Assurance, 16,* 121–132.

Reichenbach, H. (1958). *The philosophy of space and time.* New York, NY: Dover Publication.

Rossi, G. B., & Crenna, F. (2013). On ratio scales. *Measurement, 46,* 2913–2920.

SI8. (2006). *The International System of Units* (8th ed.). Paris: Bureau International des Poids et Mesures. Retrieved 19 November 2017 from www.bipm.org/en/si/ si_brochure/.

SI9 Draft. (2010, September 29). *Draft Chapter 2 for SI Brochure.* Following redefinitions of the base units. Paris: Bureau International des Poids et Mesures. (No longer available on internet)

SI9 Draft. (2013, December 16). *Draft (chs. 1, 2, 3) 9th Brochure.* Paris: Bureau International des Poids et Mesures. Retrieved 19 November 2017 from www.bipm.org/ utils/common/pdf/si_brochure_draft_ch123.pdf.

SI9 Draft. (2016, November 10). *Draft of the 9th SI Brochure.* Paris: Bureau International des Poids et Mesures. Retrieved 19 November 2017 from www.bipm.org/ utils/common/pdf/si-brochure-draft-2016b.pdf.

Simons, P. (2013). Density, angle, and other dimensional nonsense: How not to standardize quantity. In C. Svennerlind, J. Almäng, & R. Inghtorsson (Eds.), *Johanssonian investigations* (pp. 516–534). Frankfurt: Ontos Verlag.

Tal, E. (2011). How accurate is the standard second? *Philosophy of Science, 78,* 1082–1096.

VIM3. (2012). *International vocabulary of metrology: Basic and general concepts and associated terms.* JCGM 200:2012 (corrected version of the 3rd edition of 2008). Paris: Bureau International des Poids et Mesures. Retrieved 19 November 2017 from www.bipm.org/en/publications/guides/vim.html.

Wittgenstein, L. (1967). *Philosophical investigations.* Oxford: Basil Blackwell.

Index